MEDICINAL PLANTS OF INDIA

DR. V S RAMACHANDRAN
DR. ARUNA RAMACHANDRAN

notionpress.com
INDIA • SINGAPORE • MALAYSIA

ISBN 979-8-90296-758-3

Foreword

Mita Banerjee, I.F.S.

Principal Chief Conservator of Forests (Research and Education), Chennai.

Tropical forests are regarded as the world's most diversity rich ecosystem (Sanders, 2006; Sayer and Whitmore, 1991). India, being home to two of the world's bio-diverse hotspots is one of 17 mega-diverse countries, has 15 Agro-Climatic zones and medicinal plants are distributed across all bio-geographic regions, diverse habitats and landscapes. It is estimated to have about 45,000 species of plants of which over 16,000 are angiosperms/ flowering plants and Tamil Nadu with over 5500 species of flowering plants ranks first among States in this aspect in the country.

India also has one of the World's oldest medicinal traditions, which is a priceless legacy. Indian systems of medicine derive many of their curative tools from plants [Kumar *et al* (2005)] which are used as drngs. Information about these is often found in old literature *(Atharveda, Charak Samhita, Sushruta Samhita,* etc.). Plant derived medicines are widely used because they are relatively safer than the synthetic alternatives; they are easily available and cheaper [Iwu et al., (1999)]. Around 70% of India's medicinal plants are found in the tropical areas and around 30% in the temperate and alpine areas. About 6200 species of plants are estimated to be used for human and veterinary health care in the country, out of which about 2,700 plant species are reported to be used in the codified Indian Systems of Medicine.

Medicinal plants are used to maintain physical, mental, and spiritual health in all cultures and in a variety of capacities and contexts. Such plants include species with one or more part/organ that produces compounds of clinically established therapeutic value, which may be utilized directly or as precursors for drug synthesis, and others that have not been vetted as such but are believed to be therapeutically useful (e.g., as medicine, tonic, food or cultural symbol). Humans have used med icinal plants since at least 2600 BCE and perhaps for upwards of 30,000 years. Moreover, for hundreds of thousands (to millions) of years, paleolithic hominins likely used many plants to cure diseases or at least to anleliorate their symptoms. Today, an estimated 80% of people worldwide use traditional herbal medicines; they are the first line of defence against disease for tens of millions of people.

Despite the ubiquity and longstanding cultural significance of medicinal plants totalling at least 30,000 species and probably over 50,000 in the world, traditional uses of most medicinal taxa have not been evaluated clinically. In parallel, only 16% of plants thought to have therapeutic value have been tested for biological activity, although we recognize that drug discovery through synthesis has yielded significant gains in targeted pharmaceutical activity. This narrow focus of research has largely ignored the richer historical context of interactions between plants and people. Interlinked research efforts in three major disciplines, evolutionary ecology, molecular biology and biochemistry and ethno-pharmacology are poised to revolutionize the exploration and use of medicinal plants. The challenges in creating pharmaceuticals from single active compounds are being overcome by integrating evolutionary and historical perspectives on traditional herbal medicines shaped by millennia of hwnan trial and error, and linking these perspectives with novel biotechnological and ethno-botanical applications. These perspectives are opening the door to the re-interpretation of medicinal plants usage through a modern scientific lens.

Traditional knowledge and the livelihood it brings to indigenous communities are handed down over generations and threatened by commercial profiteering that reduces biocultural knowledge to commodities and privatized intellectual property rights (IPRs), that motivate wrongful filing of patent applications. Thus, protecting indigenous rights of people and countries to which they belong is essential in these engagements and should always strive for Sharing of benefits as expected by international accords especially the Nagoya Protocol, which identified clear policies around access and benefit sharing (ABS) meant to ensure the prior informed consent of indigenous communities and to guarantee their receipt of benefits.

The nature of traditional knowledge, which is often 'owned' by the community rather than by a single individual, does not fall neatly into Western legal conventions of IPRs. Recognizing the past, present and future value of plants for preventing and treating diseases is a crncial first step toward responsible stewardship of medicinal plant biodiversity for future generations. Integrative research on medicinal plants. should lead to a greater focus on conserving genetic, phenotypic, chemical, and biocultural diversity, especially near their centers of origin, where such diversity is likely to be high and preserving the complicated history and relationships between humans and medicinal plants. Plant parts of over 300 species are traded in Tamil Nadu and about 180 species have high volume consumption (i.e. requirement of more than lOOMT *I* year) constituting 80% of total trade. However, only over 20 are under regular cultivation. Over 70% of these plant drugs either from forest or 'non forest' wild involve destructive harvesting. Key challenges to conserving medicinal plants thus include biodiversity loss, degrading habitats, over-harvesting, climate change and loss of traditional knowledge.

In short, medicinal plant species are not ju st chemical factories for extraction and exploitation. Rather, they may be symbiotic partners that have shaped modern societies, improved human health, and extended human lifespans. Cultivation of medicinal plants needs to be supported preferably in a cluster mode with the required infrastructure for warehousing, drying, grading, storage and transportation so as to reduce dependence on wild harvest and thus ensme conservation in the wild. These infrastructural facilities are essential for increasing the marketability of the medicinal plants, adding value to the produce, increasing profitability and reducing losses.

I congratulate the authors for bringing out this useful book which will be beneficial to the common man in getting cured of at least the common ailments without the side-effects of mode'rn day chemical based synthetic medicines.

Website : http://ifgtb.icfre.gov.in
Fax : 91-0422-2430549

Phone : 0422 2484100
E-mail : dir_ifgtb@icfre.org

वन आनुवंशिकी एवं वृक्ष प्रजनन संस्थान
(भारतीय वानिकी अनुसंधान एवं शिक्षा परिषद)
(पर्यावरण, वन एवं जलवायु परिवर्तन मंत्रालय, भारत सरकार का एक स्वायत्त निकाय)
पि.बी.नं. 1061 कोयम्बत्तूर 641 002 तमिलनाडु, भारत

INSTITUTE OF FOREST GENETICS AND TREE BREEDING
INDIAN COUNCIL OF FORESTRY RESEARCH & EDUCATION
(An autonomous Body of Ministry of Environment & Forests, Govt. of India)
P.B. No. 1061, R.S. Puram HPO., Coimbatore - 641 002, Tamil Nadu, India

Message

Dr C Kunhikannan

I am very happy to see the compilation of information on medicinal plants which are used for curing or managing several ailments in many countries around the world. The empirical knowledge about their beneficialeffects was transmitted over the centuries within human communities. Presently people forgot the usage of very common medicinal plants around them because of either ignorance or easy availability of other medicine within their reach. If we have about 20 common medicinal plants in our courtyard, we can keep away several ailments. This compilation on the very important medicinal plants around us will help to update the knowledge on such medicinal plants. The author has also included information about very important medicinal plants which are very rare or threatened due to over exploitation or other form of threat.

I am sure that this book will be useful to medicalpractitioners who practice Ayurveda, Siddha, Homeopathy and Unani system of medicine. It will be a great help to the research scholars/students of Botany, Biochemistry, Biotechnology and other relevant groups of people, who are interested to improve their knowledge on the subject. Icongratulate the author for bringing out such important compilation on medicinal plants.

Preface

Plants are one of the most important sources of medicine and play a significant role in global health care. Medicinal plants and herbs have long been recognized as valuable sources of therapeutic and curative agents. Across the world, the use of medicinal plants has gained an important place in health systems, not only for the treatment of diseases but also for maintaining overall health and well-being. Nearly two-thirds of the world's population depends on herbal medicine for primary health care. This is mainly due to their cultural acceptance, better compatibility with the human body, easy availability, and comparatively fewer side effects.

Many modern medicines are derived either directly or indirectly from plants. Several drugs contain plant extracts, while others include active bioactive compounds obtained from plant sources. Medicinal plants can be described as plants used for the prevention and treatment of various diseases and health disorders. These plants may be wild species that grow naturally in forests, grasslands, and semi-natural ecosystems without human intervention, or domesticated species that have been developed through human selection, cultivation, and management.

Herbal medicines have formed the foundation of many traditional systems of medicine and have been used in health care practices since ancient times. Their continued importance is due to their biomedical value as well as their strong connection with cultural beliefs and traditional knowledge in many parts of the world. From 1950 to 1970, several plant-based drugs were introduced into the drug market, including deserpidine, rescinnamine, and vincristine, which were obtained from higher plants.

Recent research has further confirmed the importance of plant-derived medicines. Many useful drugs have been discovered through the study of traditional remedies, folk medicine, and the knowledge of indigenous communities. Even with the advancement of modern science and technology, traditional knowledge about medicinal plants remains highly valuable. Important drugs obtained from plants include aspirin, atropine, artemisinin, colchicine, digoxin, ephedrine, morphine, physostigmine, pilocarpine, quinine, quinidine, reserpine, taxol, tubocurarine, vincristine, and vinblastine.

Medicinal plants have offered humankind a wide range of effective drugs that help reduce or cure infections and relieve suffering caused by various diseases. Even though synthetic medicines have advanced greatly, many plant-derived drugs continue to maintain their importance and usefulness. The global use of plant-based medicines is steadily increasing due to their therapeutic value and natural origin.

Although modern medicine has made remarkable progress, there are still many diseases and infections for which suitable and effective drugs are not yet available. This has created a growing need to discover and develop safer medicines that are beneficial to both humans and the environment. Plant-based drugs have great potential in the treatment of various health problems such as inflammatory disorders, diabetes, liver diseases, and gastrointestinal disorders.

Through the recent researches on herbal plants or medicine, there have been great developments in the pharmacological evaluation of various plants used in traditional systems of medicine. Consequently, plants can be described as a major source of medicines, not only as isolated active principles to be dispensed in standardized dosage form but also as crude drugs for the population. Modern medicines and herbal medicines are complimentary being used in areas for health care program especially developing countries like India. In the present book, the authors have listed 300 medicinal plants which are found in India and this will be useful to This book is prepared as a useful guide for students and faculty of Botany, Horticulture, Forestry, AYUSH, and the general public who wish to gain a better understanding of medicinal plants. It provides valuable information on the identification, cultivation, uses, and importance of medicinal plants in health care and traditional medicine. This work will help readers explore the rich diversity of medicinal plants and understand their role in education, research, conservation, and practical applications.

Dr. V S Ramachandran
Dr. R. Aruna

Acknowledgments

I wish to express my sincere and special thanks to all my friends, well-wishers, colleagues, and mentors whose constant encouragement, valuable suggestions, and timely support helped me in bringing out this book.

First and foremost, I convey my deep sense of gratitude to the Scientists and Officers of the Botanical Survey of India, particularly Dr. M. P. Nayar, Former Director; Dr. N. C. Nair, Joint Director; Dr. A. N. Henry; Dr. V. J. Nair, my Research Supervisor; Dr. P. Daniel; and Mr. K. Vivekananthan. Their motivation, guidance, and technical support helped me in various ways during the completion of my Ph.D., preparation of research publications, securing research projects at the Botanical Survey of India, Coimbatore, and throughout my service at Kongunadu Arts and Science College and Bharathiar University, Coimbatore.

I am also thankful to my Research Associates and Junior Research Fellows, Dr. M. Mohanan, Dr. C. N. Mohanan, Dr. V. S. Raju, and Dr. A. G. Pandurangan, for their useful discussions and valuable inputs on scientific matters. I place on record my heartfelt gratitude to Dr. M. Aruchami, President; Dr. C. A. Vasuki, Secretary; Dr. Kumaraswamy, Principal (Retd.); and my former colleague Dr. S. Paulsamy for their support in undertaking research projects during my service at KASC.

I equally thank the Forest Officers, especially Dr. R. Annamalai, Mr. K. Chidambaram, Dr. T. Sekar, Dr. S. PaulRaj, Mr. V. T. Kandhasamy, Mr. Anurag Mishra, Mr. Anwardeen, Mr. C. Badrasamy, Mr. S. Anand, Mr. V. C. Rahul, Mr. Sathiyamurthy, Mr. S. Thangaraj Pannerselvam, and Mr. C. Angamuthu, along with several field staff members who accompanied and assisted me during my field visits to various forests of Tamil Nadu.

I would like to thank Mr. Darshan Shankar, Mr. Vinay Tandon, Mr. D. K. Ved, and Dr. G. S. Goraya of FRLHT, Bangalore, for selecting me as a Consulting Botanist in their MPCA programme.

I am immensely grateful to Prof. Madhav Gadgil, Centre for Ecological Sciences, Indian Institute of Science, Bangalore, whose inspiration and motivation greatly strengthened my scientific interest in the ecology and taxonomy of the Western Ghats.

I also thank Dr. Ranjit Daniel of Care Earth, Chennai, for involving me in various projects as a consulting botanist and for his continued support.

I wish to convey my sincere gratitude to Ms. Mita Banerjee, I.F.S., PCCF, Tamil Nadu Forest Department, Chennai, for kindly writing the foreword to this book. I also acknowledge Dr. C. Kunhikannan, Director, IFGTB, Coimbatore, for providing a message and appreciation for the present work.

The inspiration and encouragement received from my eldest brother, Mr. S. Brahmanandam, M.A.I.P.S. (Late), Retired Chief Post Master General, Chennai, and my sister-in-law, Mrs. B. Kannammal, are gratefully acknowledged. I also thank our family friend Mr. A. Venugopal, Kanchipuram, for his support.

Above all, I express my heartfelt gratitude to my family members, especially my wife, Mrs. A. Dhanusu, M.A., for her patience, understanding, and encouragement, which enabled me to undertake many field explorations, mostly during weekends and holidays. I also thank my daughter Dr. R. Aruna and my son Ra. Shrenivasan for their support and encouragement.

Dr. R. Aruna expresses her sincere thanks to the Management and Principal of Thiagarajar College, Madurai, for granting permission to carry out field studies in the Madurai Forest Division. She also conveys her gratitude to the Tamil Nadu Forest personnel for permitting and supporting her field explorations in Madurai and Kanchipuram districts. She is thankful to the Scientists of SACON, Coimbatore, for their guidance and support in connection with the award of her Ph.D. degree in Botany. She also expresses her heartfelt thanks to her husband, Er. D. Santhosh, and her son, SA. Aadithya, for their constant support, motivation, and encouragement in all her scientific pursuits.

We sincerely thank M/s. Sree Kumaran Computers for their valuable assistance and support in the preparation of this book.

We would like to thank Notion Press and Diksha Lohia, our Publishing Manager, for all their support in publishing the book.

About the Authors

Professor V. S. Ramachandran served as a Professor in the Department of Botany, Bharathiar University, Coimbatore. He has 35 years of experience in teaching and research. His major areas of research include angiosperm taxonomy, medicinal plants, and the conservation of endemic and threatened plant species.

He has successfully completed nine research projects funded by various agencies. He has guided 14 Ph.D. scholars and 32 M.Phil. scholars. To his credit, he has published three books as co-author, two edited books, and 127 research articles in national and international journals.

Professor Ramachandran is an elected Fellow of the Indian Association of Angiosperm Taxonomy and a Fellow of the Ethnobotanical Society. He has also visited several countries, including Australia, Singapore, Turkey, Malaysia, Thailand, Sri Lanka, and Cambodia, to present research papers at international conferences.

* * *

Dr. R. Aruna received her M.Sc. degree in Botany from Madras Christian College, Chennai, in 2006 and completed her Ph.D. in 2013 from the Department of Landscape Ecology, Salim Ali Centre for Ornithology and Natural History (SACON), Coimbatore.

She has published 20 research papers in peer-reviewed journals and book chapters. She has also participated in 30 national and international conferences, seminars, and workshops. In 2008, she was awarded the Darwin Scholarship and attended the programme held at Shrewsbury, Shropshire, United Kingdom.

Dr. Aruna has visited the United Kingdom, Sri Lanka, Malaysia, and Cambodia to present research papers at international conferences. She has also successfully completed research projects funded by the Tamil Nadu Forest Department and the University Grants Commission, New Delhi.

At present, she is working as an Assistant Professor in the Department of Botany, Thiagarajar College (Autonomous), Madurai. Her major areas of research interest include plant-bird interactions, angiosperm taxonomy, ethnobotany, citizen science, and biodiversity conservation.

Table of Contents

Introduction

India is recognized as one of the twelve mega-biodiversity countries of the world. The country is broadly divided into ten biogeographic regions, namely the Trans-Himalayan region, Himalayan region, Indian Desert, Semi-Arid zone, Western Ghats, Deccan Peninsula, Gangetic Plains, North-East India, Islands, and Coasts. This wide range of biogeographic zones has contributed greatly to the rich biological diversity of the country. Variations in physical features, climatic conditions, and ecological habitats such as forests, grasslands, wetlands, coastal and marine ecosystems, and deserts have supported an enormous diversity of plant and animal life. Although India occupies only about 2.4 per cent of the world's total land area, it contributes nearly 8 per cent of the known global biological diversity. Based on available data, India holds the tenth position in the world and fourth position in Asia in terms of plant diversity.

Indigenous and tribal communities have long possessed valuable knowledge about the uses of various plants found in tropical rainforests, including trees, herbs, shrubs, climbers, and epiphytes. Information obtained from traditional systems of medicine has played a significant role in the discovery of new plant-based products used as therapeutic agents. Plant alkaloids form one of the major active components of many Ayurvedic medicines. The medicinal value of plants is mainly due to the presence of bioactive compounds such as alkaloids, alkenyl phenols, glyco-alkaloids, flavonoids, sesquiterpene lactones, terpenoids, saponins, tannins, and phenolic esters. Several powerful modern drugs, including reserpine, tubocurarine, vincristine, aspirin, atropine, digitoxin, morphine, taxol, hyoscyamine, L-dopa, and quinine, were originally derived from plants found in forests. Traditional medicine has also contributed significantly to the treatment of malaria, as compounds such as artemisinin and quinine derivatives are important modern antimalarial drugs. More than 1,277 plant species belonging to about 160 families are reported to be used worldwide for the treatment of malaria and fever.

Modern ethnobotanical studies and drug discovery programmes continue to depend greatly on the knowledge of tribal communities and rural traditional healers. Ancient scholars such as Sushruta also emphasized the importance of proper collection time, growth stage, and age of medicinal herbs, as these factors influence their biological activity and therapeutic effectiveness. Herbal medicines are generally well accepted by the human body. A single medicinal plant may be useful in treating different ailments when used in different combinations. Such valuable information is usually available only with people who have long experience and deep knowledge of the healing properties of plants.

The demand for plant-based medicines is increasing throughout the world. This has led to the rapid growth of markets related to herbal drugs, pharmaceuticals, nutraceuticals, functional foods, and cosmeceuticals. Medicinal plants are increasingly used as raw materials for the development of new drugs because of their potential to overcome drug resistance in microorganisms. Herbal medicines, once considered part of alternative or fringe practices, are now moving towards mainstream use, as more people seek natural remedies with fewer side effects than synthetic chemicals.

In recent years, traditional medicine has gained renewed importance for several reasons. In many parts of the Western world, people are becoming more aware of both the effectiveness and side effects of synthetic drugs, leading to a growing interest in natural remedies. Modern medical care has become highly technology-oriented, and in some cases, the human aspect of care is believed to have been reduced in favour of cure. Scientific and clinical studies increasingly suggest that herbal formulations, especially polyherbal preparations used in traditional systems of medicine, have minimal or almost no side effects and may be safer than synthetic drugs or isolated single compounds. The rising cost of modern medicines, treatment failures, toxicity, severe side effects, emergence of multidrug-resistant microorganisms, and lack of effective cures for several diseases have encouraged people to consider herbal therapies as suitable alternatives.

According to estimates of the World Health Organization, nearly 80 per cent of the population in developing countries depends on traditional medicine for their primary health care needs. More than 20,000 species of higher plants are used in traditional medical practices by indigenous communities around the world. Therefore, scientific documentation of indigenous knowledge on medicinal plants is urgently needed. Such documentation will not only help preserve this knowledge from being lost but also provide useful leads for the development of new therapeutic agents.

India has a rich tradition of village-based herbal medical practices. Traditional birth attendants, bone setters, herbal healers, wandering monks, and folk practitioners have been using valuable biological resources for generations. The country also has a large number of registered practitioners in Ayurveda, Unani, Siddha, and Tibetan systems of medicine. Globally, about 5,00,000 higher plant species are known, of which around 17,000 species are found in India. Among these, nearly 3,000 species are used in ethnomedicine. About 1,500 species are used in Indian traditional systems of medicine, including nearly 600 species in Ayurveda, 500 species in Unani, and about 550 species in phytopharmaceutical industries.

A large number of Indian medicinal plants and their uses have already been documented. The Central Drug Research Institute, Lucknow, initiated a biological screening programme in 1963. Studies on a small portion of flowering plants have so far yielded around 120 therapeutic agents of known chemical structure from nearly 90 plant species. Some important drugs derived from high-altitude plants include taxol, podophyllotoxin, chiratin, allicins, valerian, artemisinin, and ephedrine. Several laboratories under the Council of Scientific and Industrial Research and many academic institutions in India are engaged in the development of new drugs from plant sources. Among the important leads obtained from high-altitude plants are antihepatotoxic compounds from Picrorhiza kurroa and tranquillizing compounds from Valeriana wallichii.

About 8,000 plant species are used in local health traditions for human health, veterinary care, agricultural applications, biopesticides, seed treatment, and biofertilizers. Around 1,800 species are systematically recorded in the codified Indian Systems of Medicine. However, only about 880 medicinal plant species are involved in all-India trade. Nearly 48 medicinal plant species are exported, while about 42 species are imported. Around 61 per cent of the traded species are collected from the wild, with little or no evidence of cultivation, while only about 10 per cent are obtained through cultivation.

This heavy dependence on wild collection has placed nearly 100 medicinal plant species under different levels of threat, requiring urgent conservation attention.

The concept of "Health for All," which is a key theme of the National Health Policy, becomes more meaningful when the requirements of veterinary medicine, soil health, and organic farming are also considered. The degradation of soil due to loss of humus and nutrients has increased the need for natural and plant-based inputs. Various expert groups and agencies have studied the demand and supply gap in medicinal and aromatic plants. NABARD has also highlighted the importance of such crops not only for human health but also for livestock, plants, and soil. Medicinal and aromatic plants have wide applications in herbal medicines, herbal cosmetics, perfumes, nutraceuticals, aromatic and spice products, biopesticides, veterinary medicines, natural dyes, food colours, non-edible oils, and oil cakes.

The Government of India has established the National Medicinal Plants Board as an apex body to coordinate activities related to the medicinal plants sector. However, further focused efforts are needed to strengthen this vast area of crop diversification, especially in view of the growing global demand. Promotion of medicinal and aromatic plant cultivation can also help generate rural employment and income. It is therefore necessary to gradually reduce the present practice of obtaining 90–95 per cent of medicinal and aromatic plant materials from forests and encourage systematic cultivation of commercially valuable species suited to different agro-climatic zones.

The cultivation and use of medicinal, aromatic, and dye plants have great potential for employment generation, particularly in rural and tribal areas. With the growing emphasis on tribal and rural development, the cultivation of medicinal and aromatic plants can play an important role in improving livelihoods and promoting sustainable development.

Medicinal plants are distributed across diverse habitats and landscapes. Nearly 80 per cent of them occur in tropical regions, particularly in different forest types of the Western and Eastern Ghats. Less than 20 per cent are found in temperate regions, high-altitude grasslands, and other ecosystems, though some of these species also possess high medicinal value. A small number of medicinal plants are found in aquatic habitats and mangroves. Field studies carried out by the senior author indicate that about 60–70 per cent of the known medicinal plants occur in dry and moist deciduous forests, compared to evergreen or temperate vegetation.

Medicinal plant species are found in several plant groups such as algae, fungi, lichens, bryophytes, pteridophytes, gymnosperms, and angiosperms. However, the present study gives detailed emphasis mainly to angiosperms. Among the 800 medicinal plant species recorded under 125 families, families such as Asteraceae, Euphorbiaceae, Lamiaceae, Fabaceae, Rubiaceae, Acanthaceae, Apiaceae, Solanaceae, and Asclepiadaceae contribute a major share. Among these, Asteraceae is represented by the highest number of species.

Climate change, habitat destruction, and the spread of invasive alien species have caused a serious decline in wild plant resources. Although modern medicines are widely used for treating various diseases,

recent experiences have highlighted the usefulness of AYUSH medicines in supporting health care, including during the COVID-19 pandemic. Recognizing the importance of traditional systems of medicine, the Central Government has increased support for research in this field. Therefore, a sound understanding of the occurrence, distribution, and uses of wild medicinal plants has become highly essential.

Keeping these aspects in view, this book has been prepared to provide useful information on medicinal plants. It will be helpful to practitioners of Ayurveda, Siddha, Homeopathy, and Unani systems of medicine. It will also benefit students and researchers from disciplines such as Botany, Biochemistry, Biotechnology, Microbiology, Horticulture, Forestry, AYUSH, and all those who wish to learn more about the importance, diversity, conservation, and utilization of medicinal plants.

Abrus precatorius L. Fabaceae

Indian Licorice, Ratti, Gundumani

Slightly woody vine, twining, attaining 3 m in length. Stems green, cylindrical, puberulent, turning dark gray, rugose, glabrous and slightly flattened when mature. Leaves alternate, pinnate, 3-5 cm long; 8-15 pairs, 0.8-1.5× 0.3-0.7 cm, oblong or oblanceolate, membranaceous, the apex rounded and mucronate, the base rounded, the margins entire; petiolules minute, slender; rachis without glands, puberulent, with a minute stipel at the base of each leaflet; petioles minute, slender, with the base slightly swollen; stipules filiform, 2-3 mm long, persistent. Inflorescences of small axillary or terminal pseudoracemes, with 5-7 flowers clustered on the swellings of the rachis; bracts minute, deciduous. Calyx campanulate, green, 3-5 mm long, puberulent; corolla pink, the standard ovate, with the center dark pink, up to 1 cm long, concave, the apex acute, the wings and keels as long as the standard, unguiculate. Legumes oblong, 2-4.5 × 1-1.5 cm, slightly inflated, with the apex elongate and recurved and the margins slightly undulate, opening along the two sutures. Seeds ellipsoid, brilliant red, with a black spot at the base, 6-8 mm long.

USES:

Seed paste is applied for skin diseases, eye diseases, rat and spider bites. Crushed leaf is applied for toothache and dental cavities. Leaf along with Sesamum seeds and jaggery are ground and eaten to stop menstrual bleeding.

Abutilon indicum (L.) Sweet Malvaceae

Country Mallow, Kanghi, Thutti

It is a perennial shrub, softly tomentose and up to 3 m in height. The leaves are ovate, acuminate, toothed, rarely subtrilobate and 1.9-2.5 cm long. The flowers are yellow in color, peduncle jointed above the middle. The petioles 3.8-7.5 cm long; stipules 9 mm long; pedicels often 2.5-5 mm long, axillary solitary, jointed very near the top; calyx 12.8 mm long, divided in to middle, lobes ovate, apiculate and corolla 2.5 cm diameter, yellow, opening in the evening. The fruits are capsule, densely pubescent, with conspicuous and horizontally spreading beaks. The seeds are 3-5 mm, reniform, tubercled or minutely stellate-hairy, black or dark brown.

USES:

Root decoction is given for UTI, lumbago and nervous debility. Root paste is applied for leprosy. Leaf is used to cure piles. Seed extract is given as a laxative.

Acacia chundra (Rottler) Brenan Mimosaceae

Brenan Indian gum Arabic, Kikai, Karungaali.

A small or medium-sized thorny tree, up to 15 m tall. Bark dark grey or greyish- brown, peeling off in long strips, or sometimes in narrow rectangular plates, brown or red inside. Branches slender, puberulous when young but glabrescent, with 2 curved, ca. 8 mm long prickles at the base of each petiole. Leaves bipinnately compound with 9-30 pairs of pinnae, and a glandular rachis; leaflets 16-50 pairs, oblong-linear, 2-6 mm long, glabrous or pubescent. Flowers in 5-10 cm long, in axillary spikes, 5-merous, white to pale yellow, with a campanulate 1-1.5 mm long calyx, and a 2.5-3 mm long corolla; stamens numerous, far exserted from the corolla with white or yellowish-white filaments. Fruit a strap-shaped pod, 5-8.5 cm × 1-1.5 cm, flat, tapering at both ends, shining brown, dehiscent, 3-10- seeded. Seeds broadly ovoid.

USES:

It is used for the treatment of cough and sore throat. The bark is said to be effective against dysentery, diarrhoea and in healing wounds.

Acacia indica (L.) Brenan Mimosaceae

Acacia nilotica (L.) Willd. ex Del. Subsp. indica. Indian Gum Arabic, Babool, Karu Velamaraam

Spreading, dense shrub or tree to 12 m high. Bark smooth, becoming rough and longitudinally fissured at base. Stipular spines mostly 2-50 mm long, often diminutive and appearing absent on some branchlets. Leaves: petiole (0.4–) 0.9-2 cm long, pubescent, mostly with a gland at or just below base of lowest pair of pinnae; rachis (0.3–) 0.8-6.8 cm long, pubescent, with a small gland at apex, rarely with a gland at base of second pair of pinnae from apex; pinnae (1–) 2-6 (–10) pairs, 1-4.5 cm long; pinnules 9-30 pairs, narrowly oblong to cultrate or slightly oblanceolate, 1.7-7 mm long, mostly 0.5-2 mm wide, obtuse to broadly rounded at apex, concolorous. Inflorescences simple, 2-6 in axils; peduncles 7-32 mm long, with involucre of bracts mostly 1/2-3/4 way from base; heads globular, 30-53-flowered, yellow to bright yellow. Mature pods narrowly and regularly constricted between seeds into orbicular sections, flat, 6-25 cm long, 14-17 mm wide (2-4 (–9) mm wide at constrictions), coriaceous, grey- or white- tomentose, sticky inside, indehiscent.

USES:

It is known for its wound healing property and widely used for oral diseases. Active constituents of the gum are Arabic acid, malic acid and tannin.

Acacia sinuata (Lour.) Merr. Mimosaceae

Soap Pod Wattle, Seekai

Woody climber. Branchlets with patchy or sparsely distributed long weak crisped hairs, with wide ribs and scattered recurved prickles up to 2 mm long along the ribs. Leaves: petiole 1.5-3.5 cm long, with a prominent elliptic gland at about the middle; rachis 5-10 cm long, with scattered abaxial prickles on some leaves, gland present at base of uppermost pair of pinnae; pinnae 5-6 pairs, 5-7 cm long; leaflets 12-24 pairs per pinna, 6.5-9 mm long, 1.8-2.5 mm wide, oblong, obtuse, mucronulate, asymmetrically truncate at base, midrib oblique (arising at leading edge of leaflet and extending to about the middle of the leaflet at the apex), 3 or 4 lateral veins with distinct reticulum between them, with sparse long loose hairs on both surfaces or hairs confined to margins. Inflorescences single or in 2's or 3's in upper axils, the subtending leaves often not developing and heads forming a raceme or panicle; head globular, 30-40-flowered; peduncles 1.5-2 cm long, indumentum similar to that of branchlets; buds red or purplish red, open flowers cream. Flowers 5merous, sessile; calyx 1.5-3 mm long, shortly lobed, glabrous; corolla 2.5-3 mm long, lobed to about level of the calyx. Fruit thick, fleshy pod, depressed between seeds, wrinkled when dry.

USES:

Leaf rasam helps to decrease the cholesterol level in the body and is recommended three times in a week. Rasam prepared from its shoot tip is recommended once in a week to increase digestion. Regular use of rasam prepared from its leaf, prevents gall bladder stones.

Acalypha alnifolia Klein ex Wild. Euphorbiaceae

Alder Leaved Cat Tail. Chinni

A densely pubescent shrub or undershrub, often woody at the base. Leaves are simple, usually alternate or spirally arranged, with a broadly ovate to elliptic shape. They are characterized by a crenate to serrate margin, an acute apex, and a cuneate base. The plant is monoecious, producing both male and female flowers on the same plant. Male flowers borne on axillary, cylindrical spikes, which can be up to 8 cm long. Female flowers located in smaller, upper axillary spikes, featuring cupular, dentate bracts that enclose the ovary. The fruit is a three-lobed capsule.

USES:

This plant is used in curing asthma, bronchitis, diabetes, dysentery, fever, hypertension, jaundice, liver inflammation and pneumonia.

Acalypha indica L. Euphorbiaceae

Indian Acalypha, Kuppaimeni

Small annual herbs, erect, at times diffuse, 30-120 cm high. Stem simple or branched, deep-green, hirsute, angular, hairy and deeply grooved in younger parts, more or less terete, glabrous and faintly straite in older. Leaves alternate, ovate, rhomboid-ovate, sometimes obovate, subacute or obtuse at apex, cuneate or tapering at base, serrate only in the upper part along the margins, dark-green above, pale-green below, glabrous or thinly hairy, petioles 3-8 cm long, hairy, stipules small. Flowers in axillary spikes. Male flowers minute, pale-greenish-yellow, condensed in the upper part, followed by a tuft of sterile flowers. Females lax, in the lower half, scattered, 3-5 leafy, large, dentate, many-nerved bracts. Sepals4, petals 0, stamens 8, filaments free, ovary 3-celled, styles filiform. Capsules 1-2 mm in diameter, green pubescent, dehiscent, concealed in the bract. Seeds 3, about 1 mm in diameter, lead-coloured or pale brown, glabrous, along ovoid, smooth.

USES:

Leaf along with salt is made into paste in lime juice, applied for ring worm, eczema and skin diseases. Leaf, garlic and pepper ground with water are used once a day for three days in case of menstrual pain.

Achyranthes aspera L. Amaranthaceae

Prickly- Chaff Flower, Naiyruvi

Erect or ascending herbs, 0.6-0.8 m high; Stems tough, becoming woody at the base. Leaves opposite, simple and ovate, up to 10 cm long by 8 cm wide, tapering to a point at both ends and shortly stalked, the blades entire. Inflorescences terminal and axillary, spicate, erect, many-flowered; becoming elongate, with only a few flowers open at the same time.; flowers hermaphrodite, solitary in axils of acute, membranous, persistent bracts. Individual flowers are small, with five white to pink or greenish tepals and white filaments, and form narrow, elongated terminal spikes up to 60 cm long. As the flowers age, they bend downwards and become pressed closely against the stem. The bracts surrounding the flowers in the fruiting stage have sharp, pointed tips making the heads spiny to the touch. Fruits are capsules, orange to reddish purple or brown, 1-3 (-5) mm long. Ovary is 1-seeded.

USES:

Root paste with rice washed water in mixed with honey and applied for piles. Leaf and pepper are ground in water are taken for bleeding piles. Leaf along with *Curcuma longa* rhizome is ground into a paste and is applied for 6-9 days to cure polyp in nose. Leaf juice is applied to stop bleeding from wounds.

Acorus calamus L. Araceae

Sweet Root, Vasambu

A strongly aromatic semi-aquatic perennial herb; rhizomes creeping, jointed, somewhat vertically compressed, 1.3-2.5 cm thick, pale to dark brown and spongy inside. Leaves narrow, up to 80 cm long, linear to narrowly ensiform, glossy bright green, apex acute, base amplexicaul; petioles sheathing for 20-50 cm. Flowers pale green, fragrant, arranged compactly on a sessile, cylindrical, stumpy spadix 5-7 cm long. Fruits (berries) green, angular, 3-celled, fleshy, containing 1-3 oblong seeds..

USES:

Rhizome is slightly tonic but forms a useful adjunct to other tonics and stimulants. It is very popular for the remedies of cough and cold and also for other respiratory disorders like bronchitis. In raw form, it is also used as cough lozenge. It also provides aid to the digestive system and acts against flatulent colic, dyspepsia, and vomiting. It depresses central nervous system, and an ingredient in formulation for psycho-somatic disorders like epilepsy.

Adenanthera pavonina L. Mimosaceae

Coral wood, Aanaikundumani

It is a medium-sized to large deciduous tree, 6-15 m tall and up to 45 cm diameter, depending on location; generally erect; bark dark brown to greyish. Leaves are very large, bipinnate with a large swollen pulvinus; 2-6 opposite or sub-opposite pairs of pinnae, each with 8-21 alternate leaflets on short stalks; leaflets 2-4.3 × 1 cm, oblong to ovate, with an asymmetric base and blunt apex, dull green above, light green beneath, turning yellow with age. Flowers in narrow spike like racemes, 12-15(-25) cm long; fragrant, small, petals 5, oblong or elliptic, cream or white lightly connate at the base, stamens 10, as long as the petals, the anthers tipped with minute glands. Legumes linear, flattened, 15-22 × 1.3-1.5(-2) cm with slight constrictions between seeds, dark brown, turning black upon ripening, leathery, dehiscent from top to bottom by twisting valves to reveal 8-12 hard-coated, vivid scarlet seeds, 7.5-9 mm in diameter, lens shaped; seeds adhere to pod.

USES:

Seed paste and honey is applied for ulcers and bruises. Bark paste is applied for septic wounds and ulcers. It is used as a medicine against rheumatism, migraines, headache and dysentery.

Aegle marmelos (L.) Correa Rutaceae

Bael, Vilvam

It is a deciduous tree, 6-8 meters in height with trifoliate aromatic leaves. The branches sometimes bear long straight spines. The bark is shallowly furrowed and corky. The bisexual flowers are nearly 2 cm wide, borne in clusters, sweet scented and greenish white. The shallow calyx has 5 short sepals and is pubescent on the outside. The 5 petals are oblong ovoid, blunt, thick, pale greenish white and dotted with oil glands. Stamens are numerous, sometimes coherent in bundles. Ovary are oblong ovoid, slightly tapering, axis wide, cells numerous (8-20), small arranged in a circle with numerous ovules in each cell. Fruits are 5-7.5 cm in diameter, globose, oblong pyriform, rind gray or yellow, pulp sweet, thick yellow, orange to brown in color. Seeds are numerous and arranged in the cells surrounded by a slimy transparent mucilage. Seeds have wooly hairs.

USES:

Leaf juice and *Saraca asoca* bark are made into a decoction and used to regulate menstrual cycle. The fruit is mildly astringent and used to cure dysentery, diarrhoea, hepatitis, tuberculosis, dyspepsia and good for heart and brain. Roots have anti diarrhoetic, antidote to snake venom, anti- inflammatory and wound healing properties.

Aerva lanata (L.) Juss. ex Schult. Amaranthaceae

Mountain knotgrass, Sirupoolai

Annual, diffuse, erect or ascending herbs, 15-50 cm tall, somewhat thickened and woody at base. Stem simple or branched glandular, hairy, striate. Leaves broadly ovate, elliptic or obovate, acute or subacute at apex, tapering at base appressedly hairy on both surfaces, entire along the margins; petioles 0.3-1.6 cm long, pubescent. Flowers in spikes, small, whitish or greenish - white, in axillary spikes on 2-4, 0.5-3.0 cm long culminating into interrupted panicles at the end of the branches due to suppression of upper leaves. Staminodes 5; filaments pale - yellow. Stigmas 2. Fruit Utricle minute, broadly ovoid, subacute, smooth, shining, 1 - seeded. Seeds black, shining, smooth, acute at one end, with margins outside.

USES:

Whole plant decoction is given with milk to the pregnant ladies thrice in a month to improve the body strength. The decoction of the root is given in a dose of 50-60 ml to treat renal calculi and retention of urine. Cold infusion of the whole plant is given in a dose of 30-40 ml to treat cough and sore throat. The paste of the plant is applied over forehead to treat headache. The paste of the root is made into decoction and given in a dose of 30 ml to treat gonorrhea.

Agave americana L. Agavaceae

Century Plant, Aanai Katthaallai

Plants acaulescent or short-stemmed, commonly suckering, trunks less than 2 m; rosettes not cespitose, 10-20 × 20-37 dm. Leaves erect, spreading to ascending, occasionally reflexed, 80-200 × 15-25 cm; blade light green to green or glaucous-gray, sometimes variegated or cross-zoned, narrowly to broadly lanceolate, smooth, rigid; margins nearly straight or undulate to crenate, armed, teeth single, 5-10 mm, 1-4 cm apart; apical spine dark brown to grayish, conical or subulate, 2-6 cm. Scape 5-9 m. Inflorescences paniculate, not bulbiferous; bracts persistent, triangular, 5-15 cm; lateral branches 15-35, horizontal to slightly ascending, comprising distal 1/3-1/2 of inflorescence, longer than 10 cm. Flowers erect, 7-10.5 cm; perianth yellow, tube funnelform to cylindric, 8-20 × 12-20 mm, limb lobes erect, subequal, 20-35 mm; stamens long-exserted; filaments inserted above mid perianth tube, erect, yellow, 6-9 cm; anthers yellow, 25-35 mm; ovary 3-4.5 cm, neck constricted, 3-6(-8) mm. Capsules short- pedicellate, oblong, 3.5-8 cm, apex beaked. Seeds 6-8 mm.

USES:

Root decoction is given for blockage in urine flow. Oil prepared from plant juice is applied for joint pain and skin diseases. Leaf poultice over stomach also has diuretic action.

Ageratum conyzoides L. Asteraceae

Billy Goat Weed, Visadodi Vaadai chedi

It is an erect, branching, annual herb reach 50-75 cm tall at flowering. The stems are covered with short, white hairs. Leaves are opposite, 20-100 mm long, 5-50 mm wide, on hairy petioles 5-75 mm long, broadly ovate, with a rounded or narrowed acute base and an acute or obtuse or sometimes acuminate tip and toothed margins, sparsely hairy on both sides. The branched, terminal or axillary inflorescence bears 4-18 flower heads arranged in showy, flat-topped clusters. Heads are light blue, white or violet, on 50-150 mm long peduncles and are 5 mm across, 4-6 mm long with 60-75 tubular florets. The bracts are 3-5 mm high, outer ones 0.5-1.75 mm wide, sparsely hairy, evenly toothed in the upper part, with an abruptly acuminate, acute tip. Flowers are 1.5-3 mm long and scarcely protrude above the bracts. The fruit is a ribbed or angled, black achene, 1.25-2 mm long, roughly hairy, with a pappus of 5, rarely 6, rough bristles, white to cream coloured, 1.5-3 mm long with upward turning spines.

USES:

Plant is rubbed over the body as an antidote for itching due to contact of *Tragia* and *Laportea*. Oil prepared using the juice of this plant is used for bums and cold. Plant juice or paste is used to stop bleeding from cuts and wounds.

Ailanthus excelsa Roxb. Simaroubaceae

Tree of Heaven, Maharukh, Perumaram

Trees, to 25 m high. Leaves pari or imparipinnate, alternate, estipulate; rachis 20-80 cm long, stout, swollen at base, pubescent; leaflets 13-29, subopposite; petiolule 20-50 mm long, slender, pubescent; lamina 9-15 × 4-6 cm, mostly ovate-lanceolate, base unequal or truncate, apex acute or acuminate, margin irregularly and coarsely dentate, membranous, pubescent; lateral nerves 12-20 pairs, pinnate, prominent, intercostae reticulate, prominent. Flowers polygamous, 8 mm across, yellowish, in subterminal panicles, emits foetid smell; male flowers: pedicels long, slender; sepals 5, ovate, pubescent; petals 5, ovate-lanceolate, reflexed; disc 10 lobed; stamens 10, inserted at the base disc; filaments shorter than anthers; anthers oblong; pistillode rudimentary; bisexual flowers: sepals, petals and disc as in male flowers; stamens 2 or 3; ovary 2-5, partite, superior, sparsely hairy, ovule 1 in each cell; style free or connate; stigmas curling. Samara, 5 × 1.3 cm, prominently veined, oblong, copper red, twisted at base.

USES:

It is used to treat wounds and skin eruptions, fevers, bronchitis, asthma, diarrhea and dysentery. The tribal population uses the plant for antifertility, anthelmintic and rejuvenating purpose.

Allium cepa L. Amaryllidaceae

Onion, Pyaz, Vengayam

It is an herbaceous biennial. The stem of the plant is a flattened disc at the base and the tubular leaves form a pseudostem where their sheaths overlap. The leaves are either erect or oblique and there are 3-8 per plant. The white flowers clustered on stalks. The bulbs are formed just above the flattened stem of the plant by overlapping leaves. The bulb is made up of several layers, each corresponding to a leaf. They are are generally oval but shape can be variable and occur in clusters of 3-18 to a plant. The bulb is protected by a membrane which turns to a papery coat. Onion plants can reach a height of 50 cm and are grown as annuals, harvested after one growing season.

USES:

It is used as a diuretic to reduce swelling. It is also help to reduce arteriosclerosis by lowering blood cholesterol levels and preventing the formation of blood clots. Onion is also used to treat diabetes and is reputed to lower blood sugar levels.

Allium sativum L. Amaryllidaceae

Garlic, Lahsun, Vellaipoondu

It is an herbaceous, annual, bulbous plant grown for its pungent, edible bulb. The plant can reach 60 cm. The bulb can be up to 7 cm in diameter and is made up of 1-15 cloves. The stem is very short and flattened and gives way to a pseudostem, The garlic plant can possess 6-12 flat, blade-like leaves which can stretch up to 50 cm long.

USES:

It is used to treat several ailments including fevers, diabetes, rheumatism, intestinal worms, colic, flatulence, facial paralysis, high blood pressure, and bronchitis.

Alpinia galanga(L.) Willd. Zingiberaceae

Greater Galangal, Barakulanjan, Perarathai

A robust, perennial herb, up to 3.5 m tall, with subterranean, creeping, copiously branched, rhizome. Rhizome subterete, 2-4 cm in diameter, hard, fibrous, shiny, light red or pale yellow, fragrant. Pseudostem erect, formed by the rolled leaf sheaths. Leaves alternate, distichous, lowest and uppermost ones smallest; sheath densely pubescent at apex; ligule truncate, 1 cm long, densely pubescent; petiole 1-1.5 cm long, hairy; blade oblong lanceolate, (20-)50(-60) cm × (4-)9(-15) cm, base cuneate, apex with short point, subglabrous, glossy green, densely white-dotted. Inflorescence terminal, erect, many flowered, racemose, 10-30 cm × 5-7 cm, pubescent. flowers fragrant, 3-4 cm long, yellow-white; calyx tubular, about 1 cm long, white; corolla tube terete, about 1 cm long, lobes 3, recurved, oblong lanceolate, 1.5 cm × 0.6 cm, margins ciliate, greenish-white; labellum (central staminode) petaloid, spatulate, 1.5-2.5 cm × 0.5-0.75 cm, white veined with lilac, undulate crenate with a stalk like base and a recurved apex; lateral staminodes represented by 2 subulate lobes at the base of the labellum, 7-8 mm long, reddish; stamen 1, erect with incurved anther, 2-2.5 cm long; style slightly longer than stamen, stigma obtriangular. Fruit a globose to ellipsoidal capsule, 1-1.5 cm in diameter, orange-red to wine red.

USES:

The rhizomes are widely used in skin diseases, respiratory diseases, as a stomachic after childbirth, for intestinal problems, cancers of mouth and stomach, and as an expectorant.

Albizia amara (Roxb.) B. Boivin Mimosaceae

Washing tree, Usil

It is a small to medium-sized, deciduous, acacia-like tree, (3-) 6-13 m high. It is many-branched and forms a wide, dense and rounded crown. The bark is thin, rough, scaly and variable in colour. Young shoots, branchlets and leaves are pubescent. Leaves are paripinnate, 10-20 cm long, bearing 10-40 pairs of pinnulae with about 15-30 pairs of leaflets. The leaflets are minute, about 2-7 mm × 0.5-2 mm, elliptic and sparsely ciliate. The flowers are white, cream, or pinkish-white in colour, and fragrant. They are grouped in showy globulous clusters, 2.5 cm in diameter. During flowering, the trees resemble cherry trees in blossom. Fruits are indehiscent, flat, straight pods. They are 10-24 cm long and 2.5-4 cm wide and greyish brown in colour. They contain 6-13 compressed, ovate, hard-coated, brown seeds.

USES:

The leaves are used in the treatment of diarrhoea, oedema and wounds. The bark is used to treat jaundice and mouth inflammations. The fruit pods are emetic. The tree yields a gum used against ulcer. Leaves used for hair wash and ophthalmia.

Albizia lebbeck (L.) Benth. Mimosaceae

East Indian Walnut, Siris, Kaattuvaagai

It is a deciduous tree, 15-20 m tall and 50 cm in diameter at maturity, with grey fissured corky bark, somewhat flaky; inner bark reddish. The leaves are bipinnate, glabrous or slightly hairy on the axis, pinnae in 2-4 pairs, each with 2-11 pairs of obliquely oblong to elliptic-oblong leaflets, 15-65 × 5-35 mm, shortly stalked. The inflorescence consists of large clusters 5-7.5 cm wide of fragrant pedunculate globular flower heads, 15-40, on stalks 5-10 cm long. The corolla is 5.5-9 mm long, glabrous, cream, white or green, with numerous pale green stamens on filaments 15-30 mm long. The entire inflorescence is 'fluffy' in appearance, 60 mm in diameter, yellow-green with a pleasant fragrance. The pods are pale straw to light brown at maturity, narrow-oblong 12-35 × 3-6 cm, papery- leathery, swollen over the seeds and not constricted between them, indehiscent and borne in large numbers. Seeds are brown, flat, orbicular or elliptic, 8-10 × 6-7 mm, transversely placed with 3-12 in each pod.

USES:

Leaf paste is applied for skin diseases. Bark decoction is given for rheumatism and is also recommended for swellings.

Aloe vera (L.) Burm. f. Liliaceae

Barbodos Aloe, Ghi Karuar, Chotthu Kattaallai

Herbs succulent. Stems short, suckering freely to form dense clumps. Leaves sub- basal, slightly distichous in seedlings and new shoots, erect, pale green, sometimes with pale spots in very young plants, linear-lanceolate, 15-35(-50) × 4-5(-7) cm, margin sparsely spiny-dentate, apex 2- or 3-dentate-pointed. Inflorescence erect, 60-90 cm; peduncle to 2 cm thick; raceme 30-40 × 5-6 cm, sometimes with 1 or 2 ascending branches, numerous flowered; bracts whitish, broadly lanceolate, 10 × 5-6 mm, veins 5-7, apex acute. Flowers reflexed; pedicel 1/2 as long as bract. Perianth pale yellow mottled with red, slightly ventricose, 2.5(-3) cm, outer lobes free for 1.8 cm, slightly recurved at apex. Stamens exserted by 4-5 mm. Style conspicuously exserted.

USES:

Leaf pulp is applied over the face to remove black head and other marks. Leaf juice along with cumin seeds and sugarcandy is recommended for blood dysentery, leucorrhoea, amenorrhoea, menstrual problems and intestinal worms. Leaf pulp mixed with egg is applied on scalp for dandruff. Leaf juice or pulp is used as shaving cream to prevent pimples and blisters.

Alpinia calcarata Rosc. Zingiberaceae

Amkolinji, Chittarathai

Pseudostems to 1.3 m. Leaves sessile; ligule 8-15 mm, glabrous, apex obtuse; leaf blade linear-lanceolate, 20-32 × 2-3.5 cm or narrower, glabrous, base attenuate, margin with well-spaced, short bristles, apex acuminate and caudate-mucronate. Panicles less than 10 cm; rachis slightly velvety; proximal branches 3- or 4-flowered; bracteoles oblong, to 1.7 cm, membranous, apex obtuse. Pedicel ca. 3 mm. Calyx to 1.2 cm, split on 1 side, pubescent, apex 3-toothed. Corolla tube white, ca. 9 mm; lobes oblong, ca. 2.2 cm. Lateral staminodes red, subulate, ca. 3 mm, adnate to base of labellum. Labellum white with rose red and purple streaks, obovate, 2.7—3.5 × 1.5-2 cm, apex emarginate. Filament ca. 1.3 cm; anther 5-7 mm. Ovary 3-4 mm in diam., sericeous. Capsule red, globose

USES:

Rhizome decoction is given for rheumatism and gas trouble. Plant decoction is used for asthma, skin diseases and poisonous bites.

Alstonia scholaris (L) R. Br Apocynaceae

Devil's Tree, Satni, Ezhilaippaalai

A tall evergreen tree upto 20 m or more; bark grey rough, lenticellate branches glabrous. Leaves (4-)5-7 (-10) whorled, coriaceous. glabrous, and shining, 9-20 × 2-5 cm, oblong-lanceolate or obvate, obtuse or shortly acuminate with 25-60 pairs of lateral nerves; margin entire; petiole 0.5-1 cm long, ending in a thick hooked gland on upper side. Inflorescence umbellately branched, many flowered, of capitate cymes, pubescent or tomentose; peduncle 2-5 cm long; pedicel 2-5 mm long, bracts oblong or lanceolate, acute, pubescent. Flowers greenish white 6-12 mm long. Calyx c. 2 mm long, pubescent, persistent; lobes oblong-ovate obtuse. Corolla tube cylindric, c. 6 mm long, lobes ovate, obtuse, 3-5 mm long. Disc absent. Follicles 30-60 cm long and 3-5 mm broad, pendulous.

USES:

Bark decoction is given for indigestion and worms. Extract of bark crushed with *Acorus calamus* rhizome and ginger is given after delivery to the ladies. Bark decoction is used as bathing water by the patients suffering from leprosy. Latex mixed with coconut milk is applied for eczema.

Alternanthera sessilis (L) R. Br. ex DC. Amaranthaceae

Dwarf copper leaf, Ponnanganni Keerai

It is an annual or perennial herb, of 0.2-1 m high, prostrate, creeping, often rooting at the nodes, cylindrical and slightly hairy, with numerous, erect branches. The leaves are simple, opposite, shortly petiolate or sessile, broadly lanceolate or spatulate to almost linear, 0.6-5 cm long, and 0.3-1 cm wide. They are attenuated at the base, and the apex is acute to blunt, with entire, glabrous or pilose (thin, fine, articulate hairs) margins. The inflorescences are dense, sessile, silvery-white clusters of compressed spikes in the leaf axils; perianth segments are equal in length, acute, 1.5-2.5 mm long with a short point. Bracts are ovate, concave, 0.3-1 mm long and persistent; bracteoles are oblong-ovate, 1-1.5 mm long, may be acute, and not deeply lacerated. Sepals are 2-3 mm long, white or purplish, glossy with a green base, glabrous or with a few long hairs, and a strong midrib. The fruits are indehiscent, a small, flattened, obcordate or obovate utricle, 2-2.5 mm long, enclosing the seed. Seeds are dark-brown to black, disc-shaped and shiny, about 0.8-1 mm in diameter.

USES:

Whole plant ground in milk is given for over menses bleeding and for dysmenorrhea. Plant decoction is recommended for rheumatism. Oil prepared by boiling leaf juice with gingelly oil is used as hair oil, for eye diseases and sleeplessness. Leaf juice mixed with gooseberry fruit juice is boiled with

Amaranthus spinosus L. Amaranthaceae

Spiny Amaranth, Mullukeerai

Much-branched erect herbs to 60-75 cm. Leaves rhombic-ovate, ovate, or ovate- lanceolate, 3-10(-15) × 1.5-6 cm, base broadly cuneate, margins entire, plane or slightly undulate, apex acute or subobtuse. Inflorescences simple or compound terminal staminate spikes and axillary subglobose mostly pistillate clusters, erect or with reflexed or nodding tips, usually green to silvery green. Bracts of pistillate flowers lanceolate to ovate- lanceolate, shorter than tepals, apex attenuate. Pistillate flowers: tepals 5, obovate- lanceolate or spatulate-lanceolate, equal or subequal, 1.2-2 mm, apex mucronate; styles erect or spreading; stigmas 3. Staminate flowers: often terminal or in proximal glomerules; tepals 5, equal or subequal, 1.7-2.5 mm; stamens 5. Utricles ovoid to subglobose, 1.5-2.5 mm, membranaceous proximally, wrinkled and spongy or inflated distally, irregularly dehiscent or indehiscent. Seeds black, lenticular or subglobose, 0.7-1 mm diameter, smooth, shiny.

USES:

Root or stem decoction is used for urinary block and other urinary disorders. Whole plant decoction is recommended for skin diseases.

Amaranthus viridis L. Amaranthaceae

Wild Amaranth, Kuppaikeerai

Erect herb, with slender stem ascending, 15-80 cm tall, glabrous to pubescent, pubescent especially upwards. Leaves glabrous or pubescent on the veins of the lower surface; petioles long (up to 10 cm), occasionally longer than the blade; blade ovate to rhombic-oblong, 2-7 × 1.5-5.5 cm, base tapered to blunt, tip rounded, minutely mucronate, barely to clearly emarginate. Flowers green, unisexual, male and female intermixed, in slender axillary to terminal paniculate spikes 2-12 cm long and 2-5 mm wide, or in dense axillary clusters in the lower part of the stem. Bracts deltoid- to lanceolate-ovate, membraneous with a short awn from the green midrib. Perianth-segments 3, about 1.5 mm long. Stigmas 2-3. Capsule nearly globose 1.25-1.75 mm long, not rupturing or rupturing irregularly, surface rough. Seed 1-1.25 mm, round, slightly compressed, dark brown to black with a paler thick border.

USES:

Whole plant decoction is used in skin diseases. Its decoction is useful for gas trouble. Plant juice mixed with jaggery is given at night for constipation. Plant extract is used as a blood purifier.

Amorphophallus paeonifolius Dennst.ex Nicolson Araceae

Elephant Foot Yam

Corm to 20 cm across, depressed globose, tubercled, pale pink inside. Petiole to 50 cm long, 3-4 cm thick, green with brown patches; lamina 70-120 cm across; lobes to 14 × 5.5 mm, ovate, acuminate, decurrent at base into a wing to the petiole; nerves impressed above. Spathe 38 cm across, campanulate, undulate, dull greenish brown; spadix as long as spathe; appendage obovoid, 20-25 cm across, uneven, rugose, deep pink. Flowers on the lower half of the spadix. Berry 15 × 10 mm, oblong, obovoid, orange-red, glabrous

USES:

The plant is a potent analgesic, anti-inflammatory, CNS depressant, anthelmintic, antibacterial, antifungal and cytotoxic agent. Used widely in curing piles.

Andrographis paniculata (Burm. f.) Wall. ex Nees Acanthaceae

King of Bitters, Creat, Nilavembu

It is an annual, branched, herbaceous plant erecting to a height of 30-110 cm in moist shady places. The stem is acutely quadrangular; much branched and can be broken easily due to its fragile texture. Leaves are simple, opposite, glabrous, lanceolate, 2-12 cm long, 1-3 cm wide with acute entire margin. Inflorescence is terminal and axillary in panicle, 10-30 mm long with small bract and short pedicel. The flower possess calyx with 5 sepals which are small and linear. Corolla tubes are narrow, about 6 mm long, bilabiate, upper lip oblong, white with a yellowish top, whereas the lower tips are broadly cuneate, 3-lobed, white with violet markings. Stamens 2, inserted in the throat, anther basally bearded. Ovary superior, 2-celled with exserted style. Capsule erect, linear-oblong, 1-2 cm long, compressed, longitudinally furrowed on broad faces with thin glandular hairs. Seeds are very small.

USES:

Plant decoction is given for fever, malaria, cough and rheumatism. Leaf decoction mixed with lime juice, long pepper powder and pepper is used for malaria. Plant paste is applied for poisonous bites. Equal quantity of whole plant along with Tinospora cordifolia, Pterocarpus santalinus and ginger are powdered, made into decoction and is consumed with sugar and honey for all kinds of fever.

Anethum graveolens L. Apiaceae

Dill, Soa, Sathakuppai

An erect herb grows upto 90 cm tall, with slender stems and alternate finely divided leaves composed of many linear segments. The leaves are borne alternately along the stalks and are sheathed at the base. The small yellow flowers are arranged in umbels (flat-topped clusters of flowers). The small dry fruits, often mistaken for the seeds, are broadly oval in shape, about 3.5 mm long, with three longitudinal dorsal ridges and two winglike lateral ridges. They are light brown in colour. The essential oil content is about 3 percent; its principal component is carvone and have a pleasant aromatic odour.

USES:

It is used in Unani medicine in colic, digestive problem and also in gripe water. Also used in used in the preparations of more than 56 ayurvedic preparations, which include Dasmoolarishtam, Dhanwanthararishtam, Mrithasanjeevani, Saraswatharishtam, Gugguluthiktaquatham, Maharasnadi kashayam, Dhanwantharam quatham.

Anisomeles malabarica (L.) R. Br. Lamiaceae

Malabar Carmint, Paeymeratti

Perennial, semi-shrubby herb; stem to about 2 m high, much branched from base, subquadrangular, thickened below to 1.8 cm in diam., densely lanate. Leaves ovate- lanceolate to oblong-lanceolate, 3.6-16 × 1.3-7 cm, narrowed and rounded at base, acute, crenate-serrate, slightly bullate and velvety lanate above, densely so beneath; petioles 0.7-3.5 cm long, stout, lanate. Racemes to 25 cm long; verticils close, dense; peduncles densely lanate. Floral leaves 8-10 mm long, densely lanate. Bracts linear, to 5 mm long. Calyx to 9 mm long; tube to 5 mm long, lanate without, glabrous within; lobes lanceolate, to 4 mm long, acuminate. Corolla 1.4-2 cm long; tube to 9 mm long, white, glabrous without; throat pilose towards base of lower lip; Upper lip oblong, 4-6 mm long, obtuse at apex, slightly arched, whitish; lower lip to 1 cm across, coral pink, with 2 white streaks towards base, pilulose without with gland-tipped hairs, the lateral lobes shallowly rounded, the median one larger, broadly orbicular, 2-fid. Style glabrous; branches linear, unequal. Nutlets ovoid, 2.5 × 1.5 mm, trigonous, glabrous, blackish-brown and shining.

USES:

Leaf juice is given for indigestion and stomachache, while decoction for digestive disorders, diarrhoea and intermittent fever.

Anogeissus latifolia (Roxb. ex DC.) Wall. ex Bedd.
Combretaceae

Axle wood, Dhaura, Vellainaga

It is a small to medium-sized tree up to 20-25 m tall, with a straight and cylindrical bole up to 80-100 cm in diameter. Its leaves are opposite or sub-opposite, lamina 2.5-18 × 1.7-9 cm, elliptic, elliptic-oblong, suborbicular, or oblong-obovate, base obtuse, round, cuneate or acute, apex obtuse, emarginate or mucronate, margin entire, glabrous, subcoriaceous; lateral nerves 4-11 paired, pinnate, prominent beneath, arched towards the margin simple with grayish-yellow or whitish hairs below. The fruit is greenish-yellow, compressed, puberulous, circular; wings 2, margin entire or slightly undulate, beaked; seed one, obovate..

USES:

The bark and gum is widely used in wound healing, skin diseases, diabetes, anaemia and haemorrhoids.

Annona muricata L. Annonaceae

Prickly Custard Apple Mullu-Sitha-Pazham

It is an evergreen tree. Bark dark brown, fissured. Leaves simple, alternate, exstipulate, coriaceous, glabrous, glaucous or slightly pubescent beneath. Flowers bisexual, terminal or leaf opposed, solitary or in fascicles, Sepals 3, valvate, Petals 3 or 6, valvate, much larger than the sepals, outer whorl fleshy and triquetrous above with a thinner concave base, inner whorl if present very small, strap shaped. Stamens numerous, anthers narrow, ovoid, dorsal at the top of connectives. Carpels many, subconnate, ovule 1, basal, erect, style oblong, carpels partly free in flower, later confluent into large syncarpous fleshy fruit, 1- ovuled. Seed with fleshy albumen.

USES:

It is used to treat and improve various diseases like Diabetes mellitus, hypertension, cancer, and bacterial infections.

Annona squamosa L. Annonaceae

Custard Apple, Sharufa/sitaophal, Seethapazham

A small tree. glabrous branches. Leaves simple, alternate, elliptic-lanceolate coriaceous., , exstipulate, oblanceolate, entire, obtuse. Inflorescence axillary, 2 or more flowers arise in axil of leaf. Flowers trimerous. Calyx made up of 3 sepals, gamosepalous, valvate. Corolla made up of 3 petals, polypetalous, valvate, pale-yellow. Stamens numerous, arranged spirally over a conical receptacle, filaments short, anthers appendaged. Stamens many with convex connective. Gynoecium polycarpellary, apocarpous, superior, spirally arranged, one ovule in each carpel, basal placentation, style short, stigma tapering and papillose. Fruit globose, furrowed, greenish-yellow, with white pulp and blackish brown seeds.

USES:

The Leaves, shoots, bark and roots have medicinal properties. It is strongly astringent and used to treat diarrhoea and dysentery. The green fruits, seeds and leaves have effective vermicidal properties. The leaves are used to lower uric acid levels in the blood. The unripe fruit is astringent. An infusion of the leaves and fruit is used to aid digestion and treat rheumatism.

Aristolochia indica L. Aristolochiaceae

Indian Birth wort, Garudakkodi

Perennial herbs or shrubs, twining or prostrate, rarely erect. Leaves entire or lobed, 3-7-nerved with a usually dilated petiole, exstipulate but often with an undeveloped axillary stipule-like leaf (pseudostipule). Inflorescence solitary, axillary or in short fascicled racemes. Perianth zygomorphic or actinomorphic, coloured, tubular, tube inflated below, hairy inside, limb oblique, (1-)2 (-3)-lipped. Stamens (5-) 6 or multiples of these with sessile anthers. Ovary 5-6-celled; placentation parietal or axile; ovules many, anatropous, biseriate; style divided into 3, 5 or 6 linear or obtuse lobes. Seeds ± compressed, usually winged all around.

USES:

Leaf decoction is given for fever, while that of root for infections after delivery, snake bite, fever and worms. Root along with the roots of Ocimum tenuiflorum and Rauvoljia serpentina are ground with lime juice and the extract is recommended for cobra bite. Root juice is used for indigestion.

Artemisia nilagirica (C. B. Clarke) Pamp. Asteraceae

Indian Worm Wood, Makkippu

Undershrubs, ca. 1-3 m tall, aromatic, perennial, fruticose; stems paniculately branched, incanus, pubescent or tomentose. Leaves simple alternate, upper ones sometimes sessile; lamina ca. 1.5-8 × 1-4 cm, ovate, lacerated or pinnatifid, lobes acute, mucronate, pubescent above, white felted below. Capitulum solitary or fascicled in paniculate racemes, globose, ca. 0.2-0.4 cm in diam., yellowish white; involucral bracts 2-3-seriate, oblong-ovate, margins scarious. Achenes minute, oblong-elipsoid, blackish; pappus absent.

USES:

This plant is used in the management of epilesy, nervous disorders, as diuretic and skin disorders. It's also known for its antibacterial, antifungal, and anti- inflammatory properties.

Asparagus racemosus Willd. Liliaceae

Climbing Asparagas, Sadawar, Thanneervittan kizhangu

Climbing perennial herb to subshrub, branches ± rectangular. Leaf spur spinescent, spine sharp, straight, 1.5-2.0 cm on the main stem and 0.5-1.0 cm on branches. Cladodes in fascicles 3-5 (-8), linear, 1-2 cm × 0.5 mm. Inflorescence axillary, many flowered raceme or panicle up to 8 cm, flowers bisexual. Bracts 1mm, broadly ovate. Pedicel 3-4 mm, articulated in the middle. Tepals 8 × 1.5 mm, white. Stamens up to 2 mm, shorter than the perianth, anthers minute. Fruits globose berry.

USES:

Tuber decoction is taken for rheumatism, leucorrhoea, burning urination, stomachache and ulcers. Roots boiled in oil are applied for septic wounds and ulcers. Tuber juice is consumed with sugar for hyperacidity. Tuber juice is mixed with sugar and milk is given as a health tonic to increase lactation, resistance, for over menstrual bleeding.

Asystasia gangetica (L.) T. Anderson Acanthaceae

Chinese violet, Miti Keerai

Erect or diffuse herbs; branches 4-angled, tomentose. Leaves opposite, 2-3.5 × 1.5-2.2 cm, broadly ovate to elliptic, base cuneate, apex acuminate, lateral nerves close, pubescent; petiole to 1 cm long. Flowers in terminal or subterminal secund racemes. Calyx lobes 5, 5-6 mm long, lanceolate. Corolla lilac; tube to 2.5 cm long, puberulous without; lobes 5, subequal, spreading, orbicular. Stamens 4; anthers c. 3 mm long. Style to 1.5 cm long, puberulous below. Capsule to 1.5 cm long, clavate, stalked, pubescent, beaked; seeds 2-4, c. 0.5 mm cross, orbicular or angular.

USES:

Plant paste is used for bone fracture shoulder dislocation. One handful plant crushed and cooked with raw rice and green gram dhal (1:1:2:) is recommended with buttermilk for 12 days in case of jaundice. Plant ground with cumin seeds is water is given for vomiting during pregnancy.

Azadirachta indica A. Juss. Meliaceae

Margosa, Vembu

It is a tree 20-35 m or higher, with a straight trunk and long spreading branches forming a broad round crown; it has rough dark brown bark with wide longitudinal fissures separated by flat ridges. The leaves are compound, imparipinnate, each comprising 5-15 leaflets. The compound leaves are themselves alternating with one another. It bears many flowered panicles, mostly in the leaf axils. The sepal are ovate and about one cm long with sweet scented white oblanceolate petals. It produces yellow drupes that are ellipsoid and glabrous, 12-20 mm long. Fruits are green, turning yellow on ripening, aromatic with garlic like odour.

USES:

Bark powder is recommended for septic wounds. Dried grapes soaked in leaf or decoction is recommended for four days in case of measles. Neem oil is applied for healing wounds and ulcer. Hot water extract of the bark is taken orally by the adult female as a tonic and emmenagouge. Dried flower, fruit and leaf is taken orally for leprosy. Hot water extract of the flower and leaf is taken orally as an anti-hysteric remedy. The dried flower is taken orally for diabetes. Hot water extract of dried fruit is used for piles and externally for skin disease and ulcers. Hot water extract of the entire plant is used as anthelmintic and purgative.

Azima tetracantha Lam. Salvadoraceae

Misteleto Berry Thorn, Kanta-gur-kamai, Mulsangu

Scrambling, deciduous, dioecious shrubs, with axillary spines, branchlets divaricate, tetragonous. Leaves simple, opposite, 3x2 cm, elliptic-obovate, spine-tipped. Flowers unisexual, small, axillary, greenish-white; male flowers in dense globose fascicles; female solitary or in 2-flowered clusters. Sepals 4, toothed. Petals 4. Stamens 4; Staminodes 4. Ovary bilocular. Fruit globose berry, small, smooth, fleshy, 5-10mm, greenish-white containing 1-2 seeds.

USES:

Plant decoction is given as uterine tonic after delivery. Plant is much valued as uterine contractor. The leaves, root, and milky juice are bitter in taste. The leaves possess diuretic action. They work on kidneys by increasing the amount of salt and water that comes out through urine. Diuretics helps in the high blood pressure, glaucoma and edema.

Bacopa monnieri (L.) Pennell Scrophulariaceae

Thyme-Leaved, Gratiola, Neerbrahmi

Herbs prostrate to decumbent, succulent, glabrous. Stems creeping, to 50 cm, loosely ascending, rooting at nodes, much-branched. Leaves sessile, 0.5-2 cm by 0.15-0.8 cm, oblong-oblanceolate, the bases cuneate, the margins entire or denticulate, the apices rounded, with 1-3 parallel veins from the base of the blade, glabrous. Inflorescences of solitary axillary flowers; pedicels 0.5 cm to 1(-3.5) cm; bracteoles 2, 1.3-4.5 mm, lanceolate-linear, below calyx. Calyx 4.2-7 mm; outer 3 sepals succulent, ovate- lanceolate, apices rounded, inner 2 lanceolate to linear, keeled, ciliate. Corolla 6-9 mm, campanulate, 5-lobed, nearly regular, but the lower central lobe somewhat longer and more separated from the other lobes, white, pink or lavender blue, often with purple lines, the throat dark purple, the tube yellow within. Stigma capitate, united. Capsule 4.5-5 mm, ovoid, enveloped in persistent calyx; seeds numerous 0.5 mm long, longitudinally reticulate.

USES:

Leaf and stem juice acts as diuretic, laxative and helps to normalize excretory system. Plant extract is a brain cell regenerator, sleep inducer and is used for biliousness. It is a well-known nervine tonic.

Balanites roxburghii Planch. Zygophyllaceae

Desert Date, Nanjundan

It is a multi-branched, spiny shrub or tree up to 10 m high. Stem with a reticulate dark brown or grey (rarely green) bark, armed with stout simple green or yellowish spines. Leaves petiolate; leaflets shortly petiolulate; lamina 2.5-6 × 1.5-4 cm., slightly asymmetric, elliptic to elliptic-obovate, apex subacute to obtuse, sometimes slightly refuse, base cuneate or rarely rounded; secondary nerves 4-6 pairs, ± prominent beneath; petiole (0.5) 0.8-2 cm. long, puberulous or glabrescent, canaliculate. Flowers in usually few-flowered sessile or shortly pedunculate fascicles; pedicels up to 1·5 cm. long, ± densely greyish-pubescent. Flowers c. 1·4 cm. in diam. Sepals 5.2-7 × 2.7-3·25 mm., ovate or ovate-lanceolate, coriaceous, caducous, densely pubescent outside, with long silky whitish hairs inside. Petals 7.2-9.5 × 2-2.4 (2.8) mm., narrowly elliptic or elliptic- oblong, rarely lanceolate- oblong, glabrous on both surfaces. Stamens with the anthers 1.7-2.2 mm. long, ovate or ovate-oblong, glabrous; filaments c. 3.75 mm. long. Ovary densely covered with long silky hairs. Drupe yellowish or green, up to 5 × 2.5 cm., usually subcylindric, more rarely narrowly ellipsoid or subobclavate, finely puberulous, sometimes glabrescent.

USES:

Fruit is used in treatment of whooping cough, skin diseases and antifertility. Leaves are used for the treatment of jaundice. In case of pain and swelling, the bark of plants is used by traditional healers. The paste of bark is prepared and applied externally on the affected part of the body to treat snake-bite and dog bite.

Baliospermum montanum (Willd.) Muell.-Arg. Euphorbiaceae

Ivory seed, Danti, Paei Aamanakku

The plant is a stout, monoecious undershrub up to 3.5 m high, with toothed leaves and stiff branches arising from the root. The upper branches bear small, lanceolate leaves, while the lower branches have large, and often broad, ovate, three- to five-lobed leaves with rounded base. Petioles are 5-15 cm long. The flowers of the plant are unisexual. In male flowers, the calyx is globose, 2.5 mm long, four to five partite, glabrous or slightly pubescent, membranous, finely mottled with a disc of six glands. Stamens are about 20 in number. Female flowers have ovate-lanceolate and pubescent sepals, and a disc about 2.5 mm in diameter. Fruit is a three-lobed capsule, about 8-13 mm long and usually hairy. Seeds are mottled, smooth, and have oily endosperm.

USES:

The roots and leaves are cathartic, pungent, thermogenic, purgative, anthelmintic, and diuretic. The roots are used in dropsy, anascara, and jaundice. Decoction of leaves is used for treating asthma. Seeds are purgative, used externally as stimulant, and are rubifacient. The oil from the seeds is hydrogogue, cathartic, and used for external application in rheumatism.

Barleria priniotis L. Acanthaceae

Porcupine Flower, Korantam, Cem-mulli

It is an erect, prickly shrub, usually single-stemmed, growing to about 0.75m tall. Spines are about 1.2 cm long. Leaves are up to 5-9 × 2.15-4 cm, elliptic, pointed, with a fine point, base wedge-shaped, sparsely puberulus, fringed with hairs on the margins, gland dotted beneath, leaf - stalk up to 2 cm. Orange-yellow flowers are borne in cymes in leaf- axils.

USES:

It is used in medicine including in treating fever, respiratory diseases, tooth- ache, joint pains and a variety of other ailments. A mouth wash made from root tissue is used to relieve tooth-ache and treat bleeding gums.

Basella alba L. Basellaceae

Country Spinach, Kodipasalai

Short-lived perennial herb up to 4(–8) m long, succulent; stem twining, slender, smooth, green or purplish. Leaves alternate, simple, fleshy; stipules absent; petiole up to 9 cm long; blade ovate to heart-shaped, 2.5-15 cm × 2-12.5 cm, usually cordate at base, acute or acuminate at apex, dark green or purplish. Inflorescence an axillary spike up to 22(–30) cm long, hanging, with long peduncle. Flowers bisexual, regular, 5-merous, sessile, 2.5-6 mm long, white, pink or purple; perianth fleshy, urceolate, remaining closed; stamens inserted near apex of perianth tube; ovary superior, 1-celled, ovoid, style with 3 linear stigmas. Fruit a subglobose pseudo-berry 4-7(–10) mm in diameter, enveloped by the fleshy perianth, purplish black, containing a violet juice, 1-seeded. Seed globose, c. 3 mm in diameter, dark brown to black..

USES:

Leaf paste with milk is applied for ulcers. Leaf is made into a paste with milk and is applied for bone swelling. Plant paste is applied for abscesses and swelling. Leaf paste with milk is applied for swelling due to urticarial rashes and bone damage. Mucilage of plant is applied on head before bath for headache and sleeplessness.

Bauhinia purpurea L. Caesalpiniaceae

Butterfly Tree, Mandarai

Medium-sized evergreen trees, 7-10 m tall. Bark grayish to dark brownish, thick, smooth; branches puberulent when young, later glabrous. Petiole 3-4 cm; leaf blade suborbicular, 10-15 × 9-14 cm, stiffly papery, abaxially almost glabrous, adaxially glabrous, primary veins 9-11, secondary and higher order veins protruding, base shallowly cordate, apex bifid to 1/3-1/2, lobes slightly acute or rarely rounded at apex. Inflorescence a raceme with few flowers, or a panicle with up to 20 flowers, axillary or terminal. Flower buds fusiform, 4- or 5-ridged, with an obtuse apex. Pedicel 7-12 mm. Calyx open as a spathe into 2 lobes, one with 2 teeth and other 3-toothed. Petals light pink, oblanceolate, 4-5 cm, clawed. Fertile stamens 3; filaments ca. as long as petals. Staminodes 5 or 6, 6-10 mm. Ovary stalked, velvety; style curved; stigma slightly enlarged, peltate. Pod linear, flat, 12-25 × 2-2.5 cm; valves woody. Seeds compressed, suborbicular, 12-15 mm in diameter.

USES:

Stem bark decoction is given for diarrhoea, ulcers, swellings, leprosy, cough and menstrual irregularities. One hand full leaf and equal quantity of rice are made into roties and are consumed in the next morning for ringworm infections.

Bauhinia tomentosa L. Caesalpiniaceae

Bell Bauhinia, Eruvatchi

Shrubs, evergreen, erect, to 4 m tall. Young branches puberulent. Stipules linear, ca. 1 cm; petiole 1.5-3 cm, slender; leaf blade suborbicular, 3-7 × 4-8 cm, papery, abaxially tomentose, adaxially glabrous, base cordate, 7-9-veined, apex bifid, lobes rounded at apex. Inflorescence a lateral raceme, 1-3-flowered; pedicel short; bracts and bracteoles linear, 4-7 mm. Flower buds fusiform, 2 cm, puberulent. Hypanthium turbinate, 5 mm. Calyx split spathaceously at anthesis. Petals light yellowish, subequal, broadly obovate, 4-5.5 × 3-4 cm, subsessile. Fertile stamens 10, unequal; filaments 1-2 cm, puberulent at base. Ovary stalked, tomentose; style slender, glabrous; stigma peltate, small. Legume flat, linear, 7-15 × 1.2-1.5 cm, sutures not ridged; valves leathery, splitting on the tree to release 6-12 seeds. Seeds brownish, compressed, 6-8 mm in diameter

USES:

Dried leaf or fresh flower extract is given for indigestion. Root bark decoction is recommended for biliousness, cough and worms. Bark powder applied for wounds. The fruit is said to be diuretic. The seed is eaten as a tonic and aphrodisiac. The leaves are an ingredient in a plaster applied to abscesses.

Begonia malabarica Lam. Begoniaceae

Malabar Begonia, Narayanasanjivi

Undershrubs, to 0.75 cm high; stem terete, glabrous. Leaves to 15 × 8 cm, ovate, strongly unequal at base, toothed and serrate along the margins, hispid; petiole to 6 cm long; stipule 13 × 5 mm, lanceolate. Peduncles to 3 cm, axillary; cymes 5-10 cm across, dichotomous. Flowers many; male sepals 2, 15 mm across, orbicular, pink; anthers 2 mm long, connective produced into a orbicular appendage; female sepals 7 × 13 mm, reniform. Capsule 2 × 1.5 cm, wings obtuse.

USES:

The leaves are used in treating respiratory tract infections, diarrhea, blood cancer and skin diseases.

Berberis tinctoria Leschen. Berberidaceae

Nilgiri Berbery, Daru Haridara, Oosikala

It is an evergreen shrubs, spiny or glabrous. Stems angular, terete or sulcate, spines are 3 or 5 fid. Leaves simple, but usually with a joint at the junction of lamina and petiole, in whorls or fascicles, attached to node and protected by spines, long shoots metamorphosed into spines, axils of short shoots bear cluster of green leaves, alternate, lanceolate, ovate, margin entire, dentate, chartaceous, coriaceous, usually petiolate. Inflorescences solitary or fascicled flowers, axillary, cymes. Flowers bisexual, sepals 6, rarely 6-12, imbricate in 2 to 4 series, bracts 2-3, appressed, caducous, petals 6, in 2 series, free, imbricate, normally with 2 basal glands towards inner side. Stamens 3-6, anther- locules opening by recurved valves, ovary simple, club shaped, ovules 1-12, subbasal, erect, style short or sessile, stigma, Fruits few seeded berry, blood red or black, ellipsoid to globose or ovoid, seeds about 2-10, reddish brown or black.

USES:

The plant is used for managing menorrhagia, reducing obesity and treating snakebites. It helps in curing various ailments, including skin diseases, wound healing, inflammation and jaundice.

Biophytum reinwardtii (Zucc.) Klotzsch Oxalidaceae

Slender tree plant, Mukkutti

Annual herb with simple stem; leaves whorled at the tip of stem, paripinnate compound; leaflets 10-20 pairs, oblong with equal base, nerves slender, oblique, uppermost pair usually largest, tip rounded; flowers yellow, in umbellate cyme carried on long slender peduncle; pedicel 5-7 mm long, usually 2-3 times as long as as calyx; sepals and pedicels covered with glandular hairs; petals yellow, almost twice as long as sepals; capsule as long or longer than fruiting calyx. The species is often confused with *B. sensivitum* in which peduncles are much more robust, pedicels hardly 1/3 as long as calyx, larger flowers and capsule shorter than fruiting calyx.

USES:

The whole plant is used to treat wounds, urinary calculi and snake bite. It has hepatoprotective, anti-cancer and neuroprotective properties.

Blepharis maderaspatensis (L.) Roth, *B. boerhavifolia* Roth Acanthaceae

Creeping Blepharis, Elumbotti

A prostrate or suberect, much branched herb roothing at nodes. Leaves (3-) 4 in a whorl, unequal, petiole 1-5 mm long; lamina thin, elliptic-ovate to oblanceolate, 2-6 (-8) × 1-2.5 (-3) cm, entire or minutely denticulate, apex acutely apiculate, puberulous above, hairy on nerves below, basally attenuate. Inflorescence usually an axillary cluster of 2-3 flowers, 1-1.5 cm long or solitary. Flowers white to purplish-pink, 1.2-1.8 cm long; bracts in 4 opposite paris, broadly spathulate-obovate 7-8 × 2-4 mm, apex obtuse with up to 4 mm long stiff, reflexed hairy, slender bristles on margins and apex; bracteoles spathulate with 2-5 mm long hairy bristles. Calyx 4-lobed, lobes unequal, ciliate margined, outer lobes 1-1.5 cm long, inner ones 7-8 mm long, all sharp-pointed. Corolla tube urceolate, c. 4 mm long, pubescent, upper lip absent, lower lip c. 1.2-1.4 cm long, obscurely lobed. Stamens 4, filaments c. 4.5 mm long, glabrous, anthers oblong, c. 3 mm long. Style c. 7 mm long. Capsule ovoid, 6-7 mm long, shining. Seeds 2, ± orbicular, flattened, c. 4 × 3.5 mm.

USES:

It is used to treat headache. Seeds are used as dysuria, diseases of nervous system, diuretic, aphrodisiac. It is used to cure cuts and wounds. The juice extracted from leaf is heated with gingelly oil and applied on affected places to heal wounds.

Blepharis integrifolia(L.f.) E. May & Drege ex Schinz

(B.repens (Vahl.) Roth.) Acanthaceae

Suberect or procumbent perennial herb, sometimes mat-forming-forming. Stems up to 40 cm long, more or less densely hairy. Leaves linear to elliptic or slightly obovate, up 6 cm long, more or less densely hairy, roughly above and on the edges; margin entire, rolled under. Flowers solitary. Bracts in 4 or 5 pairs, greenish to brown, sometimes with purple veins and blotches, oblong-elliptic to obovate, ending in a recurved bristle with 3-7 purple- red bristles, up to 3 mm long, on each side of the margin. Corolla pale blue to mauve or purple, with darker veins, 8-22 mm long; lip ovate-spathulate, hairy on both sides, 4-10 mm wide, central lobe longer and narrower than side lobes. Stamens 4, equal, filaments 1.25-1.5 mm long, thicker at the base, glabrous, prolonged above the insertion of the anther; anthers 1.5-1.75 × 0.5 mm, one- celled, margin and apex long ciliated. Ovary cylindrical, 1.5 mm long, glabrous, style 2.5-3.0 mm long, glabrous, base of the style produced downwards encircling the top of the ovary forming an inverted cup-like structure, stigma inconspicuous.

USES:

Whole plant cooked with rice and jaggery is given for one week for easy healing of bone fracture. Whole plant paste is applied to stop bleeding and for quick healing of cuts and bruises. Plant extract is given internally while its paste is applied externally in cases of damage to spinal cord. Much valued herb for the traditional bone setters. Leaf and seeds are edible.

Boerhavia diffusa L. Nyctaginaceae

Spreading Hog Weed, Mokkarattai

Prostrate or ascending herb, to 50 cm long, many-branched from a taproot; twigs cylindrical, glabrous. Leaves in unequal pairs; blades 1.2-5.5 × 1.3-4 cm, ovate to wide ovate, chartaceous, sparsely pilose, especially on veins, lower side glaucous, the apex rounded to acute, shortly apiculate, the base rounded, truncate to nearly cordate, the margins wavy, ciliate; petioles pilose, 0.5-3 cm long. Flowers nearly sessile, 2-4(-7) in terminal, subcapitate clusters on axillary racemes or terminal panicles, 10-30 cm long; the axes glabrous; bracts and bracteoles lanceolate. Calyx base 0.5-1.5 mm, puberulent, the limb funnel-shaped, red or violet, 0.6-1 mm long; stamens usually 2, slightly exserted. Anthocarp sessile, green, glandular pubescent, sticky, short club-shaped, 2-2.5 mm long, 5-ribbed.

USES:

Plant decoction is a diuretic and is used for urinary diseases. Decoction of the leaf is used to treat jaundice. The whole plant can be used as an anti- inflammatory activity. It is a cardiotonic and used to manage high blood pressure. Known to have hypoglycemic property and helpful in diabetes.

Boerhavia erecta L. Nyctaginaceae

Pandhari-Punarnara, Seemaimukkirattai

Annual herb, 5-60 cm tall, erect or decumbent at the very base, puberulous, especially in the upper part at the nodes of shoots. Leaves ovate, cordiform to elliptic, 1-3.5 cm by 0.8-2 cm, base truncate to rounded or cordate, apex acute, lower surface pale-white and occasionally with sunken red glands; petiole 1.5-3 cm long. Inflorescence axillary or/and terminal; flowers 2-7 in cymose panicles; peduncle 2-8 cm long. Flowers tubular-campanulate; pedicel 0.5-3 mm long, with 2-3 lanceolate bracteoles less than 1 mm in length. Perianth 1.5-2.5 mm long, with 5 faint ribs and a distinct constriction halfway, limb 1.5-2 mm long, white to pink. Stamens 2-3. Stigma peltate, entire. Anthocarp obconical, glabrous, 3 mm long, top truncate, with 5 ribs somewhat undulated, star-shaped in cross-section.

USES:

Plant decoction is a diuretic and is used urinary diseases. Root powder is given to stop diarrhoea. Plant decoction with *Tribulus terrestris* and *Aerva lanata* is used for burning urination. The leaves are crushed in water and the extract taken to treat diarrhoea. Sap from the leaves is squeezed into the eye to treat conjunctivitis. As a diuretic it is useful in treating cases of strangury, jaundice, enlarged spleen, gonorrhoea and other internal inflammations. In moderate doses, it is successful in the treatment of asthma.

Bombax ceiba L. Bombacaceae

Silk Cotton tree Semal, Mulilavu

It is a deciduous, small to large tree up to 35 m tall. Trunk usually unbranched up to considerable height. Bark grey, covered with hard small conical prickles. usually disappearing with age. Petiole 10-30 cm long, pulvinate at the base; stipules triangular, 5-10 mm × 4 mm with hairy margin, caducous. Leaflets 5-7, glabrous, entire, elliptic- lanceolate, acuminate, attenuate at base, more or less leathery, unequal, 5-20 cm × 2-8 cm; petiolule 1-3 cm long. Inflorescence many fascicles of 1-4 flowers borne, at or near the end of branches. Flowers large, showy, red (occasionally yellow or white); pedicel thick, 2-2.5 cm long. Calyx 3-lobed (rarely 2-lobed), cup-shaped, 3.5-4 cm long, smooth outside, densely silky within. Petals twisted in bud, stellate tomentose outside, sparcely pubescent inside, elliptic-oblong, usually recurved, 8-11 cm × 4-5 cm. Stamens c. 75, united at base in 6 phalanges, each of 11-15 stamens, the inner-most phalange surrounding the pistil is composed of 15 stamens of which 5-innermost are the largest and forked; filaments 3.5-6 cm long, pink, somewhat tomentose, flattened at base; anthers long, afterward twisted, violet. Ovary conical, green, covered with silky hairs, 0.5-1.2 cm long; style simple, 5.9-6.5 cm long; stigmas 5, filiform. 5-6 mm long. Capsule 10-12.5 cm long; oblong, woody, 5 valved, profusely to finely tomentose. Seeds brown, smooth, obovid, 6 mm long, embeded in silky white wool.

USES:

Decoction of the bark is given orally to combat fever; Decoction of the heartwood is used for diabetes; bark juice is given to reduce stomachache. The flowers are astringent and refrigerant. Plant parts like flowers, young root, gum, leaves, shoots, and bark have medicinal properties and are used as treatment for various conditions and diseases like cholera, fractures, toothache, coughs, urinary problems, influenza, and snake bites.

Boswellia serrata Roxb. Burseraceae

Salar Vellai Kunkiliyam

Deciduous trees, to 20 m high, bark yellowish-white with dark blotches, exfoliations thin, papery, smooth flakes; blaze red; exudation white gum-resin; branchlets pubescent. Leaves imparipinnate, alternate, apically clustered, estipulate; rachis 11-44 cm, slender, pubescent, swollen at base; leaflets 15-31, sessile or subsessile, opposite or subopposite; lamina 0.8-9.5 × 0.5-3.5 cm, elliptic-oblong, oblong-lanceolate, oblong- ovate, base oblique, acute, apex obtuse, margin entire or crenate, chartaceous, glabrous; lateral nerves 8-14 pairs, pinnate, faint, intercostae reticulate, faint. Flowers bisexual, small, white, in axillary or subterminal fascicled racemes; calyx pubescent, tube broadly campanulate, short; lobes 5-7, persistent; petals 5-7, 7 × 2.5-4 mm, white, ovate-oblong, shortly clawed, inflexed at apex pubescent out side except margin; disc annular, crenate, free from calyx; stamens 10, free, filaments alternately longer and shorter connective produced beyond the anther lobe; ovary sessile, superior, ovoid, 3-celled, ovules 2 in each cell; style to 3 mm, grooved; stigma 3-lobed. Fruit a drupe, ovoid, trigonous; pyrenes 3; seed 3.

USES:

The gum resin is used in the treatment of chronic lung diseases, diarrhoea, dysentery, pulmonary diseases, menorrhoea, dysmenorrhoea, gonorrhoea, syphilitic affection, piles and liver disorders.

Brassica juncea (L.) Czern. Brassicaceae

Indian Mustard, Sarson, Kadugu

Perennial herb, usually grown as an annual or biennial, up to 1 m or more tall; branches long, erect or patent; lower leaves petioled, green, sometimes with a whitish bloom, ovate to obovate, variously lobed with toothed, scalloped or frilled edges, lyrate- pinnatisect, with 1-2 lobes or leaflets on each side and a larger sparsely setose, terminal lobe; upper leaves subentire, short petioled, 30-60 mm long, 2-3.5 mm wide, constricted at intervals, sessile, attenuate into a tapering, seedless, short beak 5-10 mm long. Seeds globose, reddidh-brown.

USES:

Oil extracted from seeds is used as an application for rheumatism, pain and arthritis. Use of seeds in curries make them appetizer. Consuming seeds is useful during gastric irritation. Seed paste is applied over body parts for pain during typhoid fever, lymph node swelling, headache and mumps (it may cause burns in some cases).

Buchanania cochinchinensis (Lour.) M. R. Almeida Anacardiaceae

Cuddapah Almond, Saraparuppu

The tree is having 13-18 meter height with straight trunk, young branches clothed with silky hairs. Leaves thickly coriaceous, broadly oblong, obtuse, sometimes emarginate, glabrescent above, more or less villous beneath, reticulately veined, the nerves and veins impressed on the upper surface, base rounded, main nerves 10-20 pairs, petioles 1.2cm long. Flowers small, sessile, greenish white, in terminal and axillary pyramidal ferrugineo-pilose panicles which are shorter than the leaves, bracts small, caducous. Calyx lobes short, broadly ovate, ciliate. Petals long, ovate-oblong, subacute, disk fleshy. Stamens 10, a little shorter than the petals, filaments flattened, anthers about as long as the filaments. Ovaries 1 perfect, conical, villous, the other 4 reduced to cylindrical filaments. Drupes obliquely lentiform in 1.1.3 cm long, black, stone hard, 2 valved.

USES:

The gum from the tree is used against leprosy in traditional medicine. The roots are acrid, astringent, cooling, depurative and constipating. They are useful in the treatment of diarrhoea. The leaves are used in the treatment of skin diseases. The fruits are used in treating coughs and asthma.

Butea frondosa Roxb.ex Willd. Fabaceae

Flame of the Forest, Palasu, Purasu

Deciduous trees, to10 m high, bole crooked, irregular; bark 5-6 mm thick, grey to greyish-brown; exudation red; branchlets densely tomentose. Leaves trifoliate, alternate; stipules small, lateral, cauducous; rachis 12-20 cm long, stout, pubescent, pulvinate; stipels subulate; petiolule 5-10 mm long, stout, pubescent; lateral leaflets 8.8-13.7 × 5.5-11 cm, broadly oblong-ovate or suborbicular, base oblique, apex obtuse, terminal leaflet 11-15 × 12.5-15 cm, widely rhomboid, base obtuse, apex emarginate, silky pubescent on both sides when young, glabrous above, silky pubescent beneath when mature, margin entire, coriaceous; lateral nerves 4-8 pairs, pinnate, prominent; intercostae scalariform, prominent. Flowers bisexual, 5 cm long, bright red, in terminal or axillary, densely fascicled, racemes; calyx broadly campanulate, teeth 5, deltoid, short, upper 2 connate, velvety; corolla much exserted; petals 5, standard petal 5 × 2.5 cm, lanceolate, clawed, wings falcate 4.5 × 1.5 cm adnate to keel, keel united 4.5 × 3 cm, curved; stamens 9 + 1; vexillary stamens free; anthers uniform; ovary 2.5 cm, inferior, 1-celled, ovules2; style long, incurved, beardless; stigma small. Fruit a pod, 12.5-28 cm long, oblong, the base flat, wing-like and indehiscent, the tip splitting round the apical seed; seed obovate, compressed.

USES:

Gum powder is given for diarrhoea due to indigestion. Paste of seed powder with ghee is applied for leucoderma. Crushed flower juice is given for three days for complete sterility in females. Latex dissolved in milk is given with ghee at morning for worms, gas trouble, indigestion and acidity.

Cadaba fruticosa (L.) Druce Capparaceae

Indian Cadaba, Dabi, Vizhuthi

Rigid wiry shrubs. Leaves simple, 1.5-4 × 1-2 cm, oblong, slightly pubescent; petioles up to 0.5 cm. Racemes few flowered, greenish-white; sepals 2 + 2, 1 × 05. cm, obovate; petals 4, 1.5 cm long, clawed, to 0.4 mm, disc prominent, infundibuliformis, lateral to the gynandrophore, long hairy; stamens 4, exserted; ovary one celled, many ovuled, stigma sessile, clavate, hairy. Berry 4 × 0.5 cm, globose; seeds globose.

USES:

It is used to treat neurological diseases, cough, lung problems, constipation, eczema, and swelling.

Caesalpinia sappan L. Caesalpiniaceae

False sandal wood, Patang, Pathimukham

A small thorny tree, 5 m in height with a few prickly branches. Leaves are compound, large and abruptly bi-pinnate; with 8-12 pairs of oblong leaflets and small prickles. Flowers yellow in terminal and axillary panicles. Calyx 11 mm long, leathery, glabrous. Corolla 2 cm across, Petals orbicular, subequal, yellow the upper with a red spot at the base. Stamens waxy-white. filaments densely wooly at the base. Ovary gray - velvety. Pods are 7.5 – 10 cm long and 3.8-8 cm in diameter. obliquely oblong, polished and indehiscent and 3-4 seeded.

USES:

The heart wood is used for its anti thirst, blood purifying, anti diabetic, and to stop bleeding in gums and give firmness and strength to the gums.

Calophyllum inophyllum L. Clusiaceae

Alexandrian Laurel, Punnai

Trees 5-12 m tall. Bark gray brown or dark brown, thick, with longitudinal fissures, always exuding pellucid resins when wounded. Young shoots striate. Petiole robust, 1-2.5 cm; leaf blade shiny on both surfaces, broadly elliptic or obovate-elliptic, rarely oblong, 8-15 × 4-8 cm, thickly leathery, midvein raised abaxially, impressed adaxially, base rounded or broadly cuneate, apex rounded or emarginate. Thyrses in upper axils, 7-11-flowered, rarely shorter than 10 cm. Pedicel 1.5-4 cm. Flowers scented, white, 2-2.5 cm in diam. Sepals 4; outer 2: suborbicular, ca. 8 mm; inner 2: obovate, petaloid. Petals 4, oblanceolate to obovate, ca. 1.1 cm, concave, apex subtruncate or rounded. Ovary subglobose; stigma peltate. Mature fruit yellow, globose, ca. 2.5 cm in diam.

USES:

Bark decoction is used to wash septic wounds and ulcers. Oil extracted from the seeds is applied for rheumatoid arthritis. Bark decoction is hot in nature and recommended for dysentery. Gum obtained from the tree is applied as a wound healer.

Calotropis gigantea (L.) R. Br. Asclepiadaceae

Madar, Erukku

Large shrub up to 4m tall, much-branched at base, stems erect, up to 20 cm in diameter; bark pale grey, longitudinally cracked; young shoots woolly hairy; latex in all parts. Leaves opposite, decussate, simple and entire, sessile; stipules absent; blade broadly ovate to oblong-obovate, 9.5-20 cm × 6-12.5 cm, base cordate with semi-amplexicaul lobes, apex almost acute, short-hairy beneath. Inflorescence an axillary, umbellate to almost corymbose cyme up to 12.5 cm in diameter, peduncle 6-12 cm long, stout, secondary branches up to 2 cm long. Flowers bisexual, regular, 5-merous, white, cream, lilac or purple; pedicel 2.5-4 cm long, densely woolly hairy; calyx lobes broadly ovate, 4-6 mm × 2-3 mm; corolla 2.5-4 cm in diameter, lobes broadly triangular, 10-15 mm × 5-8 mm; corona with 5 narrow, fleshy scales, laterally compressed, 6-11 mm long, adnate to and shorter than the staminal column, forming an upturned horn with 2 rounded auricles on either side, cream or lilac to purple, with a dense longitudinal dorsal row of short white hairs; ovary superior, 2-celled, gynostegium up to 1 cm long, stigma head star-shaped. Fruit a pair of follicles, each follicle ovoid, boat-shaped, inflated, 6.5-10 cm × 3-5 cm, many-seeded. Seeds ovoid, 5-6 mm long, with 2-3 cm long coma at one end.

USES:

It is used to treat skin diseases like leprosy, eczema, digestive issues such as diarrhoea, piles, respiratory ailments and joint pain. The latex is used for its anti-tumor, analgesic, and wound-healing properties, while the root is effective against fever and parasitic infections.

Calycopteris floribunda (Roxb.) Lam. Combretaceae

Paper flower climber

Scandent shrubs; young stem densely brown pubescent. Leaves 4-9 × 2-3 cm, ovate to elliptic, base rounded, apex acute or acuminate-caudate, densely pubescent on both sides, more below, puncatate; petiole to 1 cm long. Flowers in dense axillary racemes, crowded in dense panicles at top of branches. Bracts leafy, 0.6-1.2 × 0.3-0.4 cm. Calyx tube 4-5 mm long; lobes 5, to 5 mm long elliptic cream coloured, velvety, accrescent. Petals absent. Stamens 10, 5+5 in two rows. Ovary inferior, 3-4 mm long, 1-celled; ovules 3, pendulous; stigma simple. Drupe 6-8 mm long, with 5 spreading accrescent calyx lobes. Seed single.

USES:

It is used in the treatment of diarrhoea, dysentery, jaundice and malaria. It is also have anthelmintic, anti-inflammatory, antifungal, heptoprotective and anticancerous activities.

Cannabis sativa L. Cannabinaceae

Indian Hemp, Bhang, Ganja

Erect, odorous annual herbs. Leaves long petioled, upper alternate, lower opposite, palmately dissected, 3-11 foliolate; stipules 2, free, lateral, filiform, Plants dioecious: male inflorescence fasciculate, shorter than the leaves, sub-sessile. Achene crustaceous. Seeds with unilateral endosperm, embryo large, curved; cotyledons thick and fleshy.

USES:

It has been used for the treatment of chronic pain, muscle spasms and act as anti-inflammatory, sedative and analgesic agent. It is used in the treatment of rheumatism, epilepsy, asthma, skin burns, pain, the management of sexually transmitted diseases.

Caralluma adscendens (Roxb.) R. Br. Asclepiadaceae

Kallimulayan

Succulent herbs. Stems, up to 60cm tall; Latex present. Leaves simple, small, rudimentary. Flowers 1-2 together, axillary, scattered, bisexual, 5-merous, regular, drooping, with foetid smell; pedicel 1-4 mm long Sepals triangular, 2-3 mm long, acute; corolla c. 2.5 cm in diameter, flat to shortly campanulate, pale green, often finely dotted with purple, sometimes striped, Corolla lobes lanceolate, 1-3 cm × 1-1.5 cm, bluntly acuminatevariably long-hairy at apex; outer corona bowl-shaped, lobes brownish to dark purple, inner corona lobes longer than the outer ones, deeply divided, filiform. Fruit consists of a pair of follicles, each follicle fusiform, 10-15 cm × c. 1 cm, apex acuminate. Seeds oblong, c. 12 mm × c. 4 mm, at apex with coma of white hairs 3-4 cm long.

USES:

It is mainly used for weight control; recent studies have revealed other potential benefits such as its antioxidant, energy-boosting, and glucose lowering activities. It is used in treating leprosy, rheumatism and skin conditions.

Cardiospermum halicacabum L. Sapindaceae

Baloon Vine, Kanphuti, Mudakkathaan

Climbing or trailing herbs. Leaves alternate, bi-ternate; leaflets 2-4 × 1-2.5 cm, ovate-lanceolate, deeply dentate or lobed, apex acute to acuminate, membranous; petiole to 3 cm long. Flowers white, polygamous in 3-7-flowered axillary tendril-bearing peduncles; peduncle up to 5 cm long. Sepals 4, in 2-pairs, outer pair smaller. Petals 4, white, unequal, with basal scales. Stamens 8; filaments unequal, pilose. Ovary 3-locular; ovule 1-per locule; stigma 3-fid. Fruit a globular capsule, inflated, 1.5-4 cm in diameter, 3- lobed, 3-celled, green, reddish at base or with reddish veins, papyraceous. Seeds subglobular, about 4 mm in diameter, dull-black, smooth, glabrous, hilum prominent, white, cordate, rather large.

USES:

Whole plant decoction is used for rheumatism and arthritis. Whole plant paste with cumin seeds is applied for rheumatoid arthritis and arthritis. Plant extract is daily taken for arthritis or rheumatism of ankle and knee. Oil prepared from plant juice is applied for burns, hair fall, sleeplessness and proper growth of hair. Plant paste with water is applied for joint spasm and rheumatism. Leaf decoction is used for phlegm and diarrhoea. Root boiled in oil is applied on the head for biliousness.

Carica papaya L. Caricaceae

Papaya, Papita, Pappali

Erect, fast-growing, usually unbranched tree or shrub, 7-8 m tall, with copious latex, trunk about 20 cm in diameter, soft, leaves clustered near top of plant, alternate, long- petiolate, blade suborbicular, to 80 cm long, palmately 7-11-lobed; lobes glabrous, toothed, flat; plants dioecious in nature, some monoectous cultivars; flowers aromatic, male in drooping axillary panicles to 80 cm long, with a 5-toothed green calyx and 5- toothed cream to yellow corolla; stamens 10; female flowers solitary or cymose in axils or below leaves, with 5 yellow nearly free petals to 5 cm long; ovary with 5 stigmas; fruit a large yellow to greenish-orange berry, oblong to nearly globose or pyriform, about 7.5 cm long and bitter in wild types, up to 45 cm long, with flesh 2.5-5 cm thick, sweet, juicy and of orange color in cultivars; seeds numerous in central cavity, rounded, blackish, about 0.6 cm in diameter, each enclosed in a gelatinous membrane (aril).

USES:

Fruit is a mild laxative, it controls liver functions and is useful for jaundice. Latex of young fruit is highly abortive. Latex is applied to expel worms and to heal wounds. Latex is locally applied (very carefully) for allergies. Eating fruits is beneficial in cold and rhinitis. Seed powder is applied externally for scabies with itch and ringworm. Seed powder mixed with honey is taken for convulsion due to worm infestation.

Cascabela thevetia (L.) H. Lippold Apocynaceae

Yellow Oleander, Kaner, Ponnarali

Shrub or small tree grows upto 6 m in height. Leaves are usually arranged in whorls are upto 25 cm in length, dark green in color and thin, flat and pointed. Flowers are yellow but sometimes occur in orange and white color too. Fruit is small quadrangled and green in color when young turning to brownish black-black as aging occurs, contains two- four flat seeds.

USES:

It is used traditionally for the treatment of gastrointestinal and inflammatory diseases, heart failures and skin tumors.

Cassia fistula L. Caesalpiniaceae

Indian Laburnum, Amaltus, Sarakkonnrai

Trees up to 20 m, usually glabrous. Leaves large, several-foliolate; petiole about 5 cm. long, eglandular; rachis usually 2-4 dm. long, like the petiole; stipules small, caducous; leaflets normally 4-8 pairs, 8-20× 8 cm, ovate to lanceolate, acute apically, very obtuse basally, puberulent to glabrous above, lightly pubescent below; petiolules up to 1 cm. long. Inflorescence a large, graceful, pendent, many-flowered raceme; pedicels slender, usually 3-4 cm. long. Flowers large, showy, yellow; sepals 5, usually about 6 mm. long, ovate or oblong, puberulent; petals 5, about 2 cm. long or longer, ovate-orbicular, short-clawed, venose; stamens 3-morphic, the 3 lowermost almost 3 cm. long, the anthers ovate-oblong, 4-5 mm. long, glabrous, dehiscent apically and basally; the 4 median stamens about 1 cm. long, the anther ovate-oblong, sagittate, about 4 mm. long, dehiscent from the basal lobes and apical pores; 3 uppermost stamens shorter and smaller, somewhat unequal, the anthers similar to the median ones; ovary slender, lightly pubescent. Legume reportedly cylindric, about 50 cm. long, indehiscent, with horizontal seeds.

USES:

Bark paste is applied for skin disease. Juice collected from heated fruit is taken to expel intestinal worms. Bark decoction is used for lymph node enlargement, intestinal problems, skin diseases and also as a laxative. Seed oil is given as a laxative for small children. Leaf paste is applied for alopecia totalis, leprosy, itches and urticaria. Fruit pulp is consumed by mixing with honey for asthma. Root decoction is recommended for fever and diarrhoea.

Cassytha filiformis L. Lauraceae

Dodder Laurel, Anthrakkodi

It is a perennial dextrorsely-twining (anti-clockwise, like *Cuscuta* spp.) herb forming masses of stems up to 2 m long, filiform, striate, branched, glabrous to tomentose, green, orange or bright yellow. Leaves scale-like, 1.5-2 mm long, ovate to lanceolate, apex acute. Inflorescence an axillary spike 1-6 cm long, usually solitary, 3-10-flowered; peduncle 1-3 cm long; bracts and bracteoles c. 0.6 mm long, ovate, ciliate. Receptacle glabrous. Flowers greenish-white, sessile, 1.5-2 mm long, glabrous; the three outer tepals 0.6-1 mm wide, broadly ovate-orbicular and ciliate, the three inner tepals 1.8-2.4 × 1.5-2 mm, ovate-triangular, obtuse at the apex, glabrous inside and outside, ± fleshy. Stamens 9, arranged in 3 whorls with an innermost or fourth whorl made up of staminodes; filaments of the first and second whorls c. 0.3 mm long, anthers c. 1 × 0.6 mm, dehiscing introrsely; filaments of the third whorl c. 0.2 mm long with 2 globose glands at the base, anthers c. 0.9 × 0.3 mm, dehiscing extrorsely; staminodes c. 0.4 mm long, triangular, glabrous, fleshy. Ovary 0.3 mm long, ovoid; style 0.3 mm long. Fruit a 1-seeded drupe, c. 6 × 5 mm, globose, surrounded by the glabrous accrescent receptacle, crowned with a persistent perianth; green, maturing black.

USES:

Whole plant paste with grounded ginger in butter is applied for wounds, cuts and ulcers. Oil prepared from this plant juice with gingelly oil is applied hair growth and colour. Leaf juice along with sugar is poured into eye in case of conjunctivitis or burning in eyes. Whole plant paste is applied over the head for alopecia areata. Plant decoction or extract in milk is used internally for leucorrhoea, over menstruasl discharge, rheumatism and also as hepatic tonic in children. Plant paste is applied over the head for insanity and mental problem.

Catharanthus pusilla (Murr.) G. Don Apocynaceae

Tiny Periwinkle, Sadabahar, Paalaich Chetthai

It is a small erect annual herb, grows up to30 cm tall. The stem is quadrangular. Leaves opposite, ovate-lanceolate 3-8 cm long with rough margin, base narrows to a tiny stalk. The whole plant contains milky sap. Small white, periwinkle like flowers appear singly or in pairs, in the upper leaf axils. Flowers are tiny, 6-8mm, flat, with 5 petals. Seed pod is very slender, 3-5 cm long, straight, pointed.

USES:

Plant is used in cancer and diabetes; root paste is used in septic wounds; root decoction is used in fever; leaves are used in menorrhagia.

Catharanthus roseus (L.) G. Don Apocynaceae

Madagascar, periwinkle, Nithyakalyani

Perennial subshrub, woody at base, 30-60 cm tall, young branches pubescent. Leaves oblong or oblanceolate, membranous, 2.5-6 × 1.5-3 cm, entire, obtuse or mucronate, base obtuse or cuneate, glabrous or sparingly hairy specially on nerves beneath, shining green above, paler below, lateral nerves obscure; petiole short, 3-10 mm long, with many glands at the axil. Inflorescence of axillary 1-4 flowered cymes, Flowers pink or white, pedicel 1-3 mm long. Calyx tube short, lobes linear acute, 5-7 mm long, hairy. Corolla tube 20-25 mm long, sparsely pubescent above, throat densely hairy within below the stamens, lobes oblong-rounded, or obovate, 15-25 mm long spreading Disc scales higher than ovary, c. 2-3 mm long. Ovary pubescent; style c. 2.5 cm long; stigma pentagonal. Follicles 2, slender, cylindric, striated. 15-25 mm long.

USES:

Root decoction is recommended to lower the high blood pressure. Root bark or whole plant decoction hypotensive and sedative. Flower or leaf extraction taken for diabetes and high blood pressure. Two spoon of crushed leaf extract in water is used twice a day for stomachache due to worms in small children.

Catunaregam spinosa (Thunb.) Tirveng. Rubiaceae

Mountain pomegranate, Maniphal, Madkarai

Small trees with small straight axillary spines. Leaves opposite on short lateral branchlets, to 4.5 × 2 cm, obovate, obtuse, tomentose below, petiolate; stipule ovate, cuspidate. Flowers solitary, terminal on lateral branches, pedicellate; calyx tube 5 mm long, lobes obovate, hispid; corolla tube 6 mm long, broad, densely villous at the base inside; lobes 5, 12 mm long, obovate, twisted, white; stamens 5, anthers sessile at the mouth of the corolla; ovary 2-6-celled; ovules many; style 12 mm long, stout; stigma fusiform, ribbed. Fruit an obovoid berry, 4 × 3 cm, glabrous; seeds many, embedded in pulp.

USES:

The fruit have useful emetic, diaphoretic and antispasmodic properties. It is useful in cases of acute bronchitis and asthma and it is applied externally in fever. The bark is abortifacient, astringent, sedative and analgesic. It is given both internally and externally in the treatment of fevers.

Cedrus deodara (Roxb.) G. Don. Pinaceae

Himalayan Cedar, Devadaru

It is evergreen conifer which can grow upto 35m. The bark is blackish-brown with closely spaced ridges and cracks, and twigs are brown and slightly hairy. It has a distinct shape, with several trunks and clear horizontal layers in its structure. Grow up to 30-40 m (occasionally 60m) tall with spicy-resinous scented wood, thick ridged or square- cracked bark, and broad, level branches. The shoots are dimorphic and are made up of long shoots, which form the framework of the branches. And short shoots, which carry most of the leaves.

USES:

The wood acts as an expectorant and is useful in curing piles, epilepsy, stones in the kidney and bladder, useful in fevers. The oil is antiseptic and helpful in curing skin diseases, wounds, urogenital diseases, diaphoretic as well as an insecticide.

Celastrus paniculatus Willd. Celasteraceae

Black Oil Plant, Malkankani, Valuluvaii

Twiners with pale-brown, warty-lenticellate bark. Leaves alternate, broadly ovate, obovate, elliptic-oblong or nearly orbicular, glabrous and serrated in upper part. Flowers greenish- yellow, unisexual in terminal or axillary racemes. Calyx pubescent outside, lobes semiorbicular, ciliate. Petals oblong, rounded at the apex. Male flowers: Stamens inserted on the margin of the disc; filaments short, anthers oblong, rudimentary ovary small, subconical, slightly 3-toothed at the apex. Female Flowers: Ovary globose, narrowed into a short stout style, glabrous; stigma large, trilobed. Stamens inserted on the edge of the disc which is larger than the disc in the male flowers; anthers small, without pollen, ovate, acute or subtriangular. Capsule, subglobose, bright yellow, transversely wrinkled.

USES:

It is used as brain tonic for increasing memory. Seeds and oil both are effective in forgetfulness and memory disorders. In dementia, it stops its progress by preventing cell damage in the brain.

Centella asiatica (L.) Urban Apiaceae

Indian Pennywort, Brahma manduki, Vallaarai

Perennial creeping herb, rooting at the nodes, but sometimes forming a large taproot. Stem terete, with shallow grooves, sometimes purplish. Leaves solitary or in groups of up to 5, pubescent or glabrous; lamina 1-7 cm. wide, reniform to almost orbicular, with a deep basal sinus; margin crenate, glabrous or with scattered hairs on upper part of petiole (petiole sometimes densely hairy when young). Umbels subcapitate; peduncles 1·5-5 cm. long, glabrous or pubescent, usually much shorter than the subtending petiole. Flowers 2-8, inconspicuous, hermaphrodite, with an involucre of 2 ovate, membranous, persistent bracts. Pedicels slender or obsolete; petals 5, dark crimson to greenish-white, 1 mm, orbicular with a slender inflexed point. Calyx teeth and stylopodium obsolete; styles short, divergent. Fruit 3·5 × 3 mm., orbicular to ellipsoid, brown at maturity, deeply constricted at the commissure and flattened laterally, primary ribs prominent when ripe, secondary ribs evident; carpophore entire, splitting in two at maturity with a seed in each half.

USES:

Plant juice is taken internally for liver disorders and mouth ulcers. Plant juice mixed with tender coconut water is given to drink for eczema and erysipelas. Plant is chewed along with cumin seeds for mouth ulcers. Plant extract is consumed daily to improve circulation, neurological problems, gastrointestinal system and as a brain tonic. Plant juice is used for fever in children and also serve as a blood purifier and diuretic.

Citrullus colocynthis (L.) Schrad Cucurbitaceae

Bitter apple fruit, Lndrayan, Paeykomatti

Perennial trailing herbs with somewhat woody tuberous root. Stems shortly hairy when young becoming scabrid. Tendrils simple, rarely bifid, slender, glabrous. Leaf blade elongate-ovate in outline, 10-60 (-110) mm long, 8-55 (-66) mm broad, distinctly scabrid-hairy beneath, smooth except on the nerves above, palmately deeply 3-5-lobed, ultimate lobes pinnately lobulate with central lobe longest, long-ovate in outline; petiolate, petiole rather densely rough-hairy; probracts lanceolate-elliptic, 4-5 mm long and c. l. 5 mm broad, caducous. Male flowers on long pedicels; calyx campanulate, ca. 9 (4-8) mm long, lobes ca. 5 mm broad; corolla pale yellow, ovate-acute, ca. 8 mm long and ca. 5 mm broad; female flowers on longer pedicels than male, receptacle-tube short, lobes lanceolate, ca. 5 mm long. Ovary hairy, subglobose or obovate. Fruit on long stalk, c. 5 cm in diameter, globose, smooth, longitudinally green striped, 5-7 cm in diameter, eipcarp thin, filled with a dry spongy very bitter pulp. Seeds numerous, ovate-oblong, yellowish brown, not margined, ca. 6 mm long and ca. 3 mm broad.

USES:

The fruits are used in diabetes, leprosy, common cold, cough, asthma, bronchitis, jaundice, joint pain, cancer, toothache, wound, mastitis, and in gastrointestinal disorders such as indigestion, constipation, dysentery, gastroenteritis, colic pain.

Cissampelos pareira L. var. hirsuta (Ham.ex DC.) Forman Menispermaceae

Velvet leaf Pareira, Urikkakodi

Climbing herb; branchlets densely tomentose. Leaves alternate, peltate, suborbiular-reniform, 2.7-3.3 × 4.3-5, cordate at base, entire at margin, retuse at apex, tomentose; 5-nerved at base; petiole 4.5 cm long. Male flowers; subcorymbose, solitary, axillary small. Female flowers; racemes, solitary, axillary; bracts foliaceous, ovate, 1.5 cm long. Flowers ca. 02 mm across, greenish. Male flowers; sepals 4, obovate, 1.6 mm long, pilose. Petals 4, connate, 1.4 mm long, hispid. Stamens 4; filament connate; anther ovoid, transversely dehiscing. Femal flowers; seplas 4, largely obovate, 1.6 mm long. Petals 4, obovate, 0.8 mm long. Ovary obovate, sessile; 1-loculed; ovules 1; style short; stigma 3- fid. Drupes obovoid, 04 mm across, endocarp curved.

USES:

Root decoction is used for curing high blood pressure, malaria, diabetes and as a blood purifier. It is also used to treat rheumatism and joint inflammation. Leaves are used for managing asthama, bronchitis, cough and fever.

Cissus quadrangularis L. Vitaceae

Bone Setter, Hodjod, Pirandai

Rambling shrubs, with 4-angled stem, constricted at the nodes; tendrils stout Leaves alternate, broadly ovate-rhomboid to reniform, 3-5.5 × 5.5-3, truncate-retuse at base, crenate-dendate at margin, subacute-obtuse at apex, glabrous; ca. 4-5-pairs; petiole ca. 2 cm long. Stipules ovate, deciduous. Umbelliform axillary, 2.5 cm long. Flowers ca. 02 mm long, pinkish-white; peduncle slender, 2.5 cm long. Calyx minutely 4- lobed, ca. 01 mm long. Petals 4, ovate-oblong, ca. 1.3 mm long, acute at apex. Stamens 4; filament filiform, ca. 02 mm long; anther ovoid. Ovary globose-ovoid, 01 mm long, glabrous; 2-loculed; ovules 2 per cell; style subulate, ca. 1.1 mm long; stigma small. Berries globose, 06 mm long, red; seeds1-2, ovoid-obovoid, 04 mm across, glabrous.

USES:

About 30cm long stem piece is cooked with rice or its decoction is given for rheumatoid arthritis (Over dose may cause skin rashes or allergy). Plant decoction is used for backache and burning sensation in feet. Stem cooked with rice is eaten for osteoporosis. Dried plant or young stem powder is taken for digestive disorders. Plant decoction is recommended for joint pain as it has pain relieving property. Plant paste is used for bone setting in case of bone fracture. It is also used for ankle twist and nerve spasm.

Cleoma viscosa L. Capparaceae

Asian spider flower, Hurhur, Naaikadugu

Erect viscid herb upto 45 cm, clothed with glandular hairs. Leaves 3-5-foliolate, petiolate; leaflets obovate, elliptic-oblong, very variable in size, often 2-4 cm long, 1.5-2.5 cm broad, middle one largest; petiole up to 5 cm long. Racemes elongated, up to 30 cm long, with corymbose flowers at the top and elongated mature fruits below, bracteate. Flowers 10-15 mm across, whitish or yellowish; pedicels 6-20 mm long; bracts foliaceous. Sepals oblong-lanceolate, 3-4 (-6) mm long, 1-2 mm wide, glandular-pubescent. Petals 8-15 mm long, 2-4 mm broad, oblong-obovate. Stamens 10-12 (rarely more, up to 20), not exceeding the petals; gynophore absent. Fruit 30-75 mm long, 3-5 mm broad, linear- oblong, erect, obliquely striated, tapering at both ends, glandular-pubescent, slender; style 2-5 mm long; seeds many, 1-1.4 mm in diam., glabrous with longitudinal striations and transverse ridges, dark brown.

USES:

Decoction prepared from its seed is used for fever. Leaf paste is applied for wounds and cuts for easy healing. Seed decoction is taken for diarrhoea and gas trouble. Whole plant paste is applied externally for wound, ulcers and skin diseases. Seed paste is applied for rheumatic pains.

Clerodendrum inerme (L.) Gaertn. Verbenaceae

Glory Tree, Ganiyari, Sangun Kuppi

Straggling shrub with foetid smell up to 2.5 m high; well-branched. Leaves opposite, ovate-obovate, 2-4 × 1.5-2, cuneate at base, entire at margin, retuse-obtuse, shortly apiculate at apex, glabrous; 5-pairs beneath; petiole 06 mm long. Flowers ca. 3.5 cm long, white, solitary, axillary, 2-3-flowered; peduncle, 3 mm long; pedicel, up to 09 mm long. Calyx campanulate, 06 mm across, lobe 5, broadly triangular, acute at apex. Corolla tubular-salvarform, tube slender, 3 cm long, lobe 5, suborbicular-obovate, 06 mm across. Stamens 4, equal, united with corolla tube at meddle; filament filiform, 2.4 cm long, whitish-pink, longer then the corolla, pubescent at base; anther oblong, 02 mm long. Ovary obovate-oblong, 4-lobed, 02 mm long, truncate at base and apex; 4-loculed imperfectely; ovules 1; style slender-filiform, 3.1 cm long, whitish-pink; stigma deeply 2-lobed, 1.5 mm long, acute at apex. Drupes globose-obovoid, 2 cm long, 4-lobed; seeds 1, obovoid, 1.2 × 1 cm long, glabrous.

USES:

Root paste is applied for poisonous bites. Oil prepared using its root is used for rheumatism and arthritis. Whole plant decoction is recommended for rheumatism. Leaves can be used as a mosquito repellent.

Clerodendrum phlomidis L.f. Verbenaceae

Sage glorybower, Agnimanth, Thazhuthaazhai

A large bush or small tree, reaching 5 m. high, with more or less pubescent branching branches. Leaves 3.8-6.3 by 3.2-3.8 cm., ovate or subrhomboid, obtuse or acute, coarsely crenate-dentate or subentire, undulate, glabrous above, more or less puberulous beneath, base truneate or subcordate; petioles 6-20 mm. long. Flowers fragrant, in small dichotomous axillary cymes. Calyx 1cm. long or more, divided about half way down, glabrous, not enlarged in fruit, segments ovate, acutely auminate, veined. Corolla white or pinkish; tube 2-2.5cm long, slightly pubescent outside, glabrous inside; lobes nearly eqal, exceeding 6 mm long, elliptic, obtuse, veined. Filament slightly pubescent below. Ovary and style glabrous. Drupe 6mm. long, broadly obovoid, depressed, the top about level with the points of the persistent calyx-lobes, normally 4-lobed with 1 pyrene in each lobe.

USES:

The decoction of the root is aromatic, astringent and used as a demulcent in gonorrhoea. The juice of leaves is used as an alternative and bitter tonic. The plant is reported to be useful in inflammation, glycosuria, pox, coryza, scrotal enlargement and postnatal complaints.

Coccinia grandis (L.) Voight. Cucurbitaceae

Ivy guard, Bimb, Kovai

Perennial dioecious climber, 5 – 25 m, with glabrous, angled stem, becoming wood with papery bark. Leaves simple, variable in shape, 5-9 × 4-9 cm broadly ovate- cordate, shallowly to deeply palmately 3-5 lobed, basal sinus wide. Flowers white, star- shaped, solitary, axillary. Calyx campanulate; lobes 5. Corolla campanulate, shortly 5- lobed. Stamens 5, arranged in three groups, monothecous and extrorse. Gynoecium is tricarpellary, syncarpous, ovary inferior, unilocular, parietal placentation, style short, stigma 3, forked and feathery. The fruit is a smooth, bright red, ovoid to ellipsoid berry, 5-7.1 cm long.

USES:

Oil prepared from leaf juice is applied over the head as a cooling agent to induce sleep in sleeplessness. Leaf juice mixed with butter is applied for herpes and burning bruises. Leaf juice is applied over the neck and forehead for sleeplessness, nervous debility, headache and giddiness. Stem or root juice is given for diabetes.

Cocos nucifera L. Arecaceae

Coconut Palm, Narial, Thennai

Tall annulate palm, thickened at the base. Leaves pinnatisect; leaflets numerous. Linear-leanceolate; petioles very stout. Spadix interfoliar, simple panicled, with flexuous branches; branches bear scattered female flower s, often between two male flowers towards their bases and male only above. Spathe woody, boat-shaped. Male flowers: small; sepals 3, petals 3, stamens 6. Female flowers: larger, ovoid; perianth greatly accrescent; sepals 3; petals 3; ovary 3-celled. Fruit ovoid, terete or trigonous, 1-seeded drupe; pericarp thick fibrous; endocarp stony with 3 pores. Seed cohering with the endocarp; endosperm fleshly, lining the endocarp with a large cavity filled with sweetish fluid.

USES:

Ripe coconut water is smeared all over the body before bath for four days to get relief from prickly heat. Inflorescence decoction is given to women suffering from infertility and menstrual irregularities. Tender coconut husk juice is drunk and is applied externally for all types of skin diseases due to increased body heat.

Codariocalyx motorius (Hont) H. Ohashi Fabaceae

Telegraph Plant, Namaskari, Thozhuganni

A semi-herbaceous and somewhat variable undershrub, usually 4-5 ft hig; stems terete, indistinctly striate, green, glabrous, usually unbranch. Leaves normally 3 foliate; petiole 0.5-1.5 in. long; stipules 0.3-0.6 in. long, subulate with a broad base, erect, striate. Terminal leafets 1.5-4 by 0.3-1.5 in., oblong or oblong lanceolate, obtuse at both ends, quite entire, rather membranous, almost glabrous and often with irrigular. glaucous blotches above, glaucous and more or less densely adpressed-pubescent beneath; main lateral nerves 7-10 on either half, oblique, arched, very slender. Lateral leaflets when present 0.25-0.7 by 0.05-0.12 in., rotating by jerks under atmospheric stimulus; petiolules 0.06-0.15 in. lng, slender stipeles about the same length, very slender. racemes axillary and terminal, the lateral often branced, 26 in. long; pedicels 0.2-0.3 in. long, finely downy; bracts about 0.25 in., caducous, ovate-cuspidate, concealing the flower buds. calyx 0.1 in. long, campanulate; Corolla 0.25-0.4 in. long, dull purple. Stamens diadelphous. Pod 1-1.7 by 0.2 in., falcate, sub-glabrous or glabrescent.

USES:

In Siddha medicine, it's used as an antidote, cardiac tonic, and for wound healing. The extracts from leaves and flowers have been used as a home remedy for treating wounds.

Coix lacryma-jobi L. Poaceae

Job's Tears, Gurlu, Nelpavalam

Perennial, robust, monoecious, herb; nodes thick, glabrous. Leaves convolute, flat, spinulosely serrate, midrib thick, glabrous. Recemes axillary and terminal; lower spikelets solitary, female; male spikelets protrude through the apex of female spikelets; peduncle very long. Female spikelets enclosed in a hardened, polished involucre. Glumes ovate, margins incurved; lemmas ovate; palea ovate-oblong; styles basally united. Fruit broadly ovate, oblong caryopsis, furrowed in the middle, white.

USES:

Fruit extract is taken to expel phlegm. Root decoction is recommended for urinary disorders and menstrual problems.

Coldenia procumbens L. Boraginaceae

Creeping Coldenia, Tripungkee, Seruppadai

Trailing herbs, stem and leaves softly white villous. Leaves 1.5-3 × 0.5-1.5 cm, obliquely obovate or subpinnatifid, base cuneate, margin irregularly dentate, apex emarginate, lateral veins impressed above. Inflorescence axillary few-flowered sessile cymes or some times solitary. Calyx 4-lobed; lobes ovate-acute, ciliate. Corolla white, ca. 2 mm long, tube ca. 1 mm long; lobes 4, imbricate in buds. Stamens 4, included. Ovary 4- lobed, 4-locular; ovule 1-per locule. Drupe of 4 subconnate pyrenes, 1.5-2 mm across, subglobose.

USES:

It is used in medicine for ailments like rheumatic swellings, immature abscesses, and leucorrhoea. The recent studies indicating its potential anti- cancer properties for treating liver cancer. Modern pharmacological studies have also revealed for its anti-hyperlipidemic and hepatoprotective activities.

Commelina benghalensis L. Commelinaceae

Kanchara, Kunavazhai

Diffuse herb, with creeping branches, sometimes develop leafless underground branches bearing cleistogamous, white, fertile flowers. Leaves simple, ovate or oblong, base contracted into petiole, often asymmetrical, pubescent; sheaths with ciliate margins. Spathes 1-3 together auricled on one side, funnel-shaped, pubescent. Aerial flowers blue. Sepals 3, membranous. Petals 3, clawed, unequal, two larger, Stamens 3 fertile; staminodes 3. Fruit pyriform capsule.

USES:

Whole plant powder is consumed for constipation. It is also used to treat sore throats, jaundice.

Commiphora wightii (Arnott) Bhandari Burseraceae

Indian Bdellium tree, Guggul, Maishakshi

It is much-branched, dioecious, trees up to 6 m tall with brown coloured, spine scented knotty, crooked and spirally ascending branches ending in sharp spines. Bark shiny, ash to yellowish white coming off in rough flakes exposing the greenish underbark, which also peels off in thin papery rolls. Leaves small, sessile, rhomboid- (ob) ovate, 1-3 leaflets, highly aromatic, leathery, shinning green on top and greyish below with irregularly toothed edges. Flowers small, unisexual, sessile, brownish red, occurring singly or in groups of 2-3, 8-10 lobed disc and an oblong-ovoid ovary; stamen 8-10. Fruit an ovoid green berry like drupe, reddish, 6-8 mm in diameter. Seed generally contain an under developed embryo.

USES:

It's known for its anti-inflammatory, antioxidant, and immunomodulatory properties. Vastly used for treating umpteen health crises, including obesity, arthritis, indigestion.

Corallocarpus epigaeus (Rottl.) C. B. Clarke Cucurbitaceae

Redfruit Creeper, Sataampanja, Akasa Garudan Kizhangu

Vines. Leaves 3-5 angled, to 8 × 8 cm, pubescent, apex acute, base cordate, dentate; petiole to 3 cm. Flowers monoecious, yellow; male flowers in racemes, to 1 cm; peduncles to 5 cm, stout; corolla 2 mm across, pubescent; stamens 3, erect; female flowers solitary, 2 mm across; ovary 4 mm, oblong, beaked. Fruit 1.5 × 1 cm, ellipsoid, beaked, circumscissile at base; seeds ca. 8.

USES:

It is used i n medicine in curing acute dysentery, venereal diseases, skin diseases and snake bite.

Cordia myxa L. Boraginaceae

Sebastian Plum, Lasora, Naruvili

Deciduous trees 3-5 m tall. Young shoots and branches with a rusty pubescence, glabrate at length. Leaves 6-12.5 × 43-8.2 cm, suborbicular. elliptic-ovate to oblong-ovate, 3-nerved, acute or obtuse, subentire to sinuate-crenate or dentate, glabrous to dense tomentose on under surface, base cuneate to rounded. Petiole 2.5-43 cm long.. Flowers in terminal lax corymbose cymes. Calyx campanulate, 5-lobed, silky-pubescent within. Corolla companulate, white, 5-lobed. Stamens 5, exerted. Drupe 20 mm long, ovoid, apiculate, brownish-yellow, base partly surrounded by the enlarged, broadly cupular calyx.

USES:

Seed paste with coconut oil is applied for ringworm. Bark juice mixed with coconut milk is given for stomachache. Ripe fruit juice is swallowed for dry cough, stomachache and headache.

Coriandrum sativum L. Apiaceae

Coriander, Dhaniya, Kotthamalli

It is an erect annual herb. The leaves of the plant are variable in shape, broadly lobed at the base of the plant, and slender and feathery higher on the flowering stems. It is a soft, hairless plant. The flowers are produced in small umbels and are white or very pale pink in color with the petals pointing away from the centre of the umbel longer than those pointing towards it. The plant produces an oval shaped fruit which is yellow-brown in color and contains two seeds.

USES:

The fruits are used as an analgesic, carminative, digestive, anti-rheumatic and antispasmodic agent.

Costus speciosus (Koenig) J. E. Smith Zingiberaceae

Kushtha, Venkostam

An erect succulent plant, up to 1.5 meters high; root stock tuberous; stem sub- woody at the base; Leaves 15-30 × 5.7-7.5 cm sub sessile, oblong, spirally arranged, silky- pubescent beneath; sheaths coriaceous; flowers in very dense spikes, many; bracts ovate, pubescent, bright red; bracteoles keeled, ciliate on the margins; corolla tube short, lobes ovate-oblong subequal; lip white with yellow center, crisped, concave; disk with a tuft of hair at the base. Stamens yellow at apex. Fruits capsule, globosely trigonus, red; seeds black with white aril.

USES:

Cooked tuber is given along with food to dogs suffering from rabies and indigestion. Plant extract is used for whooping cough. Rhizome decoction is given for albumin in urine and menstrual disorders. Rhizome juice is given by adding sugarcandy power for phlegm. Rhizome juice boiled with gingelly oil is applied for ringworm in dogs.

Couroupita guianensis Aubleat Lecythidaceae

Cannon Ball Tree, Topegola, Naagalingam

It is a big deciduous tree. Leaves simple, alternate crowded towards the ends of the branches, 8-30 × 3-10 cm oblong-ovate, base cuneate and apex acute to acumi-nate, margin entire with 15-25 pairs of secondary veins. Inflorescences Cauline, born on thick, tangled extrusions that grow on the trunk, below the foliage branches. The extrusions are from 60-180 cm long. The inflorescences are unbranched racemes, sometimes branched and paniculate. Pedicel/ hypanthium 1-6 cm long. Some trees flower profusely, until the entire trunk is buried in flowers. Sepals 6. Petals 6, broad, yellow or red outside, pink inside. Stamens many, basally connate on ring and extended on one side into a curved androphore. Fruit globose, cannon-ball-like, hard outside, brown, having foetid smell when ripe.

USES:

Oil prepared by boiling its leaf juice with gingelly oil is applied for hair fall. Stembark decoctions used for curing hypertension, tumours and inflammatory.

Crataeva magna (Lour.) DC. Capparaceae

Ovoid Berried Lingam, Barun, Mavalingam

Trees, to 10 m high, bark 2-3 mm thick, surface grey, smooth, longitudinally wrinkled. Leaves alternate, digitately trifoliate; rachis 4-12 cm long, slender, grooved above, glabrous; petiolule 4-7 mm long, slender; leaflets 10-20 × 3.5-6 cm, unequal, ovate, ovate-lanceolate or broadly elliptic, base acute, or subacute, apex acute or acuminate, margin entire, glabrous, coriaceous; lateral nerves 9-15 pairs, slender, parallel, prominent, intercostae reticulate, faint. Flowers bisexual, creamy white, in terminal corymbs; pedicel to 5 cm long. Sepals 4, 3 mm long, free, oblong, adnate to the lobed disc. Petals 4, ca. 2 cm long, creamy white, often tinged with purple, long-clawed, obovate. Disc incurved, nectariferous. Stamens many, free, longer than the petals, inserted at the base of gynophore; anthers basifixed, filaments spreading. Gynophore 4-5 cm long at anthesis, slender; ovary superior, ellipsoid, 1-[2]-celled, ovules many; stigma sessile, capitate. Fruit a berry, globose or ovoid, yellowish-grey, rind woody, scurfy, 4-5 × 2.5-3.5 cm; seeds brown, embedded in pulp.

USES:

Bark and leaf decoction is used as a blood purifier. Bark paste by boiling with oil is applied for rheumatism. Bark decoction is used for rheumatism and allergy. Bark paste is applied for body pain after a fever attack.

Crossandra infundibuliformis (L.) Nees Acanthaceae

Orange Nail Dye, Priyadarsha, Kanakkambaram

A small glabrous, evergreen undershrub, up to 1 m tall with terete branchlets. Leaves crowded near the spike, on 1-4 cm long petiole; lamina ovate-elliptic to lanceolate- oblong, 5-10 × 2-5 cm, basally attenuate, glabrous, margin undulate-crenate, acute at tips. Spikes dense, terminal and axillary, up to 15 cm long. Flowers bright orange- scarlet with yellow throat; bracts lanceolate, 12-15 × 4-5 mm, mucronate, glandular- pubescent to glabrescent; bracteoles linear-lanceolate, hairy, mucronate. Calyx lobes hairy, lanceolate, 7-9 mm long, outer 2 obtuse or bifid, inner smaller. Corolla tube c. 2.5 cm long, limb c. 1.5 cm long. Stamens inserted at the middle of tube. Ovary 2.5-3 mm long; style filiform, up to 2 cm long, glabrous. Capsule oblong, 10-12 mm long, glabrous. Seeds densely clothed with fringed scales.

USES:

Seed decoction is used as an aphrodisiac agent. Leaf paste applied for toothache. Bark ground with turmeric and the paste applied to skin diseases of children. Crushed fruits made into a paste for brushing teeth to cure pyorrhea. Root paste applied around vagina to cure venereal diseases and syphilis. Flowers ground with pepper and the paste is applied to wounds.

Cucumis madraspatana L. Cucurbitaceae
Mukia maderaspatana (L.) M. Roem

Madras pea pumpkin, Aganaki, Bilari, Musumusukai

Prostrate or climbing scabrid herbs; tendrils simple. Leaves 4-8 × 3-7 cm, ovate–deltoid, angular or shallowly 3-5-lobed, base cordate, margin denticulate, apex acuminate, mucronate, scabrid on both sides; petiole to 6 cm. Male flowers in axillary, sessile clusters. Calyx-tube to 2 mm, villous; lobes subulate, erect. Petals 5, ca. 3 mm long, ovate-oblong, obtuse, yellow. Stamens 3 free, inserted at base of calyx-tube; anthers oblong, ciliate. Female flowers solitary or in clusters. Ovary villous. Berry ca. 1.2 cm across, globose, red. Seeds lenticular, rugose.

USES:

The leaves are used for treating bronchial asthma, allergic bronchitis, chronic bronchitis, bronchiectasis, productive cough and cold, upper and lower respiratory tract infections and difficulty in breathing.

Curculigo orchioides

Gaertn. Hypoxidaceaelndian musuli, kali moosli,

It is a herbaceous tuberous perennial with a short or elongate root stock bearing several fleshy lateral roots; Leaves sessile or petiolate 15-45 × l.3-2.5 cm, linear or linear lanceolate, tips sometimes rooting, scape very short, clavate; Flowers in racemes, distichious, yellow, lowest in the racemes 2 sexual, perianth segments elliptic, oblong, hairy on the back; Fruits capsules, derived from inferior tricarpellary, syncarpous ovary, 1-4 seeded; Seeds black, oblong, deeply grooved in wavy lines.

USES:

Tuber decoction is used as a general health tonic. Tuber decoction has aphrodisiac, sperm count increasing, cooling, diaphoretic properties and is given for leucorrhoea, fever, asthama, rheumatism, bronchitis and biliousness.

Curcuma aromatica Salish. Zingiberaceae

Wild Turmeric, Kachur, Kasthuri-manjal

Pseudostem with large herbs 90-120 cm high; rhizome palmately branched aromatic tubers, cream coloured inside. Leaves in basal tufts, 65 × 20 cm, lanceolateoblong, base deltoid, pubescent below, Flowers pink, fragrant, in lateral condensed spikes; bracts ovate, recurved; fertile, pale green tinged with red or pink. Calyx unequally 3-lobed. Corolla funnel-shaped; lip yellow, obovate, deflexed, obscurely 3-lobed. Lateral staminodes obtuse. Fruit a globose capsule.

USES:

Rhizome decoction is used internally for cough, cold and bronchitis. Rhizome paste is applied externally as a cosmetic, also for curing skin diseases and wounds.

Curcuma domestica (Medik.) Valeton *(Curcuma longa* L.) Zingiberaceae

Turmeric, Haldhi, Manjal

Tall herb; rhizome aromatic, bright, bright-yellow within. Leaves in basal tufts, oblong-lanceolate, tapering to the base. Infroescence appearing with the leaves and central to the leaf-tuft; coma bracts white and green, tinged with pink; fertile bracts pale-yellow, funnel-shaped; lobes subequal. Lateral staminodes oblong; labellum obovate, with a central yellow band. Fruit ovoid capsule. The main rhizome (ground turmeric) is ovate or pear-shaped, up to 4 cm. long and 3. cm thick. The upper part is encircled by leaf-scars, the lower part is marked by scars of the secondary rhizomes and roots. It is sliced before drying. The secondary rhizomes (long turmeric) are 0.5-1.5 cm. thick, elongated, indistinctly ringed, simple or sparingly branched.

USES:

Crushed fresh rhizome boiled with milk is taken as an effective remedy at the starting stage of viral fever. Turmeric and pepper decoction is used for sore throat. Turmeric powder slightly heated with castor oil is applied for heel cracks. Turmeric paste with *Eclipta prostrata* juice is applied for urticarial and rashes. Turmeric ground with *Aloe vera* is applied to the tip of breast in case of block in milk flow.

Cuscuta reflexa Roxb. Convolvulaceae

Pea-Fruited Dodder, Amarbel, Kodiyar Koondal

Slender parasitic herb, with twining, filiform, yellow stem 1-2 mm thick often for froming a tangled mass. Leaves reduced to minute scales. Flowers small, up to 10 mm long, white, in fascicles. Calyx 5-lobed; lobes triangular-ovate, with a tubercular keel on the back. Corolla yellow, campanulate; lobes 5, triangular-ovate, spreading and the tips inflexed; with a ring of fimbriate scales at the base of filaments. Stamens 5. The capsule is elongated, conical, with a pair of simple stigmas on a very short single style. The seeds are large, 3-3.5 mm long.

USES:

Whole plant paste is applied for scabies. Plant extract with milk is taken internally as a cooling agent, useful for rheumatisum, menstrual irregularities and leucorrhoea. Oil prepared from plant juice is applied for increasing the length of hair and black colour. Mucilage extracted from whole plant is applied over the head for biliousness.

Cyathula prostrata (L.) Blume Amaranthaceae

Small Prickly Chaff Flower, Lal Chirchita, Cherukadaladi

An annual to perennial herb, 30-50 cm tall, erect or ascending, rooting at the nodes. Leaves opposite, simple, rhomboid-obovate to rhomboid- oblong, 1.3-15 cm × 0.7-6.5 cm, base contracted or narrowed, rounded, apex triangular, acute to obtuse, entire, ciliate, margin or blade often tinged red; petiole short. Inflorescence an erect, elongated raceme, terminal and in highest leaf-axils, straight or sinuous, 18-30 cm long, rachis densely pubescent; peduncle 1-12 cm long; flowers in small clusters. Flowers small; tepals 5, free, in bisexual flowers ovate-oblong, 2.5-3 mm long, strongly mucronate, dull pale green, glabrous within, externally clothed with patent long, white hairs, tepals of sterile flowers 1. 7-2.5 mm long, sessile; stamens 5, filaments connate at base, 1 mm long, anthers 2-celled, ; ovary superior, obovoid, I-celled, funicle short, style filiform, stigma capitellate. Fruit an ellipsoid utricle, 1.5-2 mm long, thin-walled, glabrous, I-seeded, surrounded by stiffperianth. Seed ovoid or ellipsoidal, 1-1.5 mm long, shiny brown.

USES:

Leaf or Stem juice is applied to arrest bleeding from cuts. Matured stem is used as a tooth brush. Whole plant decoction with cumin seeds is given for vomiting, diarrhoea and amoebic dysentery.

Cycas circinalis L. Cycadaceae

Sago fern Palm, Jangli madhan, Eentha Panai

A small, evergreen palm-like tree, upto 5m. Dioecious trees, clothed with woody bases of the petioles. Leaves in terminal crowns, pinnate; leaflets linear, acuminate at tip. Microsporophylls densely aggregated to form large terminal cones. Megasporophylls loosely arranged, crowded round the apex of stem; Ovuliferous scales loosely imbricate round the apex of stem, leaf-like, pinnatifid with 1-3 pairs of ovules along the margins of the stalk. Seeds ellipsoid to globose, compressed. As a slow growing plant, the seed can take from 6-18 months to germinate.

USES:

Pith powder is a source of starch which is used as food for digestive disorders.

Cyclea peltata Hook. f. &Thoms. Menispermaceaa

Indian Moon Seed, Pathi, Paadai Kizhangu

It is a scandent or climbing shrubs, dioecious. Leaves simple, alternate, non- peltate, or peltate, palmately veined, petiole normally long. Inflorescences cauliflorous, racemose or thyrsoid paniculated, axillary, extra-axillary, terminal or on old stems. Flowers unisexual. Male flowers, sepals 4-5 or 6, connate or rarely free, 4-5 lobed, petals 4 or rarely 5, free or connate into a cup, rarely none, stamens 4-5, connate into a peltate synandrium, anthers transversely dehisced. Female flowers sepals 1-2, petals 1-2 rarely none, carpels 1, style short, stigma trifid. Fruits drupes, obovoid, curved, with 6 rows of tubercles, endocarp horse shoe shaped. Seeds curved, cotyledons slender, narrow embryo, terete, endospermic.

USES:

Tuberous roots are also used to treat jaundice, stomach ache, fever and asthma.

Cynodon dactylon (L.) Pers. Poaceae

Bermuda grass, Dud, Arugam-pullu

Hardy, perennial grass, very variable, with long rapid-growing, creeping runner or stolons, rooting at nodes, forming a dense tuft on the surface of the soil, runners sometimes 20 m long; leaves 2.5-20 cm long, 2-6 mm broad, flat or sometimes folded or convolute; inflorescence on culms 15 cm to 1 m tall consisting of 2-12 spikes arranged star- like at apex of stem; spikes 2.5-10 cm long with numerous spikelets, arranged in 2 rows on one side of spike; spikelets flat, 2-2.5 mm long, awnless, with 1 floret; glumes unequal, the upper longer and one-third to three-fourths length of floret.

USES:

The whole plant is used in antiseptic, astringent, diuretic, cough, cramps, tumours, urogenital disorders, warts, and wounds. The plant is bitter, sharp hot taste, good odour, laxative, brain and heart tonic and emmenagogue.

Cyperus rotundus L. Cyperaceae

Coco grasss, Motha, Korai

It is a perennial herb, 7--40 cm tall with fibrous roots and reproduces largely by rhizome and tubers. Flowering stems are erect, up to 60 cm tall, 3-sided, smooth with swollen bases (basal bulbs). The leaves have a distinct midrib, are linear, usually shorter than the flowering stem, up to 7 mm wide and emerge from a sheath around the shoot base. The inflorescence is a terminal, open umbel subtended by several leafy bracts. Several unequal rays, 2-6 cm long, support 3-8 reddish-brown to purplish-brown, flattened spikelets, 1-2 cm long and 2 mm wide, each with up to 30 glumes, 3.5-4 mm long. The fruit is a 3-angled achene, 1.5 mm long, dark brown or black.

USES:

Rhizomes are used for the treatment of fever, diarrhoea, jaundice, erysipelas, gout, alcoholism, loose teeth, cough, vomiting, thirst, epilepsy and eye diseases.

Dalbergia latifolia Roxb. Fabaceae

East Indian Rosewood, Kala sheeshan, Itti

Deciduous or evergreen medium-sized to large tree up to 40 m tall; bole straight or slightly twisted, branchless for up to 12(–24) m, up to 80(–150) cm in diameter, often with prominent buttresses; bark surface whitish to grey, thin, becoming flaking; crown rounded to dome-shaped. Leaves arranged spirally, imparipinnately compound with (3–)5-7(–9) leaflets; stipules small, caducous; petiole and rachis glabrous; petiolules up to 1 cm long; leaflets alternate, broadly obovate to elliptical-oblong, 4-12 cm × 2.5-9 cm, obtuse, rounded or notched at apex, papery or thinly leathery, glabrous. Inflorescence a terminal or axillary panicle 5-15 cm long, laxly branched, almost glabrous, many- flowered. Flowers bisexual, papilionaceous, 6-8 mm long, distinctly pedicellate; calyx campanulate, c. 4 mm long, lobes shorter than tube, lower lobe longest, upper lobes fused; corolla whitish, with obovate standard and clawed wings and keel; stamens usually 9, fused into a tube, but free in upper part; ovary superior, with distinct stipe at base, style short. Fruit a flat, elliptical to oblong, papery pod 4-10 cm × 1.5-2.5 cm, with stipe up to 1 cm long, glabrous, reticulately veined, indehiscent, 1-3(–4)-seeded. Seeds kidney-shaped, 7-10 mm long.

USES:

Seed oil is applied for septic wound and ulcers. Bark decoction is recommended for urinary disorders, rheumatism and also as a wound healer.

Datura innoxia Mill. Solanaceae

Downy Thorn Apple, Safted datura, Vellai Umathai

It is a stout, annual to perennial herb up to 1 m high. Stems and leaves tomentose, covered with short and soft greyish hairs that give the whole plant a greyish appearance. Mature leaves broadly ovate, the lamina up to 20 cm long, almost entire, slightly sinuate, or irregularly lobed towards base, with conspicuous pinnate venation, alternate. Calyx 5-11 cm long, 3-6-lobed; lobes 13-20 mm long, sometimes incompletely separated. Corolla 12-19 cm long, white with green veins; limb undulate, appearing 10-lobed, alternate lobes broadly triangular or ending in a slender point 5-10 mm long. Stamens not exserted; anthers 8-10 mm long. Style 10-14 cm long; stigma well below anthers. The fruit is a globose or ovoid spiny capsule with numerous slender spines, about 3-5 cm in diameter, deflexed, spiny; spines numerous, slender, sharp, all about equal in length, to 10 mm long; persistent base of calyx to 20 mm long, very prominent. Capsule split when ripe, releasing brown seeds, 4-5 mm long.

USES:

It is used in traditional Ayurveda and indigenous medicine to treat respiratory issues like asthma, bronchitis, rheumatism, arthritis, and skin conditions like ulcers, wounds due to its antispasmodic and analgesic alkaloids like scopolamine.

Datura metal L. Solanceae

White Thorn Asiatic Apple, Safted datura, Karu Umathai

Stout annual herb, branching above, glabrous or sparsely pubescent with simple hairs. Leaves ovate to broadly lanceolate, in juvenile phase large and coarse, later leaves often 6-12 cm long, 3-8 cm wide, margins entire to sinuate dentate, apex acute to acuminate, base oblique, petioles 2-5 cm long. Flowers solitary in stem forks, peduncles 5-10 mm long; calyx angularly tubular, 5-lobed, ca. 5 cm long, the lobes triangular, 1-1.5 cm long; corolla single, double, or triple, white, purple, or yellowish, up to ca. 15(-20) cm long, the lobes terminated by a pointed tip 1 cm long; stamens 5 or more; anthers 10 mm long; ovary ca. 6 mm long; style up to 10 cm long; stigma laterally flattened. Capsules globose to ovate, ca. 5 cm in diameter, inclined, tuberculate or with short conical spines, opening irregularly, persistent calyx base somewhat thickened. Seeds numerous, yellowish brown, flattened, obovoid, 4 mm long, 3 mm wide.

USES:

Paste of its leaf, leaf of *Tinospora cordifolia* and gingelly oil cooked in cow milk is applied for herpes. Boiled leaf juice is applied for mumps. Paste prepared by mixing its leaf juice with jaggery is applied for swellings.

Decalepis hamiltonii Wight & Arn. Asclepidaceae

Sarasaparilla swallow-Wort, Mahali kizhangu

Twining shrubs; branchlets jointed. Leaves to 6 × 4.5 cm, obovate-elliptic or orbicular, apex obtuse, base cuneate, membranous; petiole to 1.5 cm. Cymes trichotomously branched; calyx deeply 5 lobed, 2 mm, oblong; corolla tube 1 mm, lobes 3 × 2 mm, oblong, recurved, white pubescent inside; corona of 10 scales, alternately long hooked and short; filaments 1 mm, anthers attached to style apex; ovaries 1 mm.

USES:

It is used in curing various diseases like stomach disorders, gastric ulcers and to stimulate appetite. The roots used as blood purifier and appetite stimulant. It is used as a food and in health drinks.

Delonix elata (L.) Gamble Caesalpiniaceae

Creamy Peacock Flower, Safed Gulmohor, Vaadanaayaanaan

An erect tree, 6-10 m tall, bark ash-coloured. Leaves 6-20 cm long, bipinnately compound, pinnae 4-8 pairs, 4-6 cm long, leaflets l0-20 pairs, sessile or subsessile, linear- oblong, obtuse, sometimes slightly mucronate, narrowed towards the base, 7-15 mm long, c. 2-3 mm wide. Inflorescence terminal raceme, the lower flower with the longest pedicel, pedicel hairy, up to 2.5 (-3) cm long. Calyx 2-3 cm long, fused at the base, leathery, silky pubescent, oblong, narrow and sharp pointed. Corolla white in the beginning, turning yellow, the upper petal is smaller and darker in colour; sub-orbicular, curled at the margins, c. 2.5 cm long. Stamens hairy, dark coloured, 5-10 cm long, thickened at the base. Pods 12.5-20 cm long, c. 6-25 mm broad, smooth, narrow at both the ends. Seeds 4-8.

USES:

Leaf paste is applied for rheumatic joints. Bark decoction is given for rheumatism.

Delonix regia (Hook.) Raf. Caesalpiniaceae

Flamboyant Tree, Gulmohor, Mayil Kondrai

Small, diffuse or erect, suffruticose undershrubs with branches finely tomentose; stems ca. 0.3-1.2 m. Leaves compound, alternate, spiral; rachis ca. 4-10 cm long, slender, with a sessile discoid gland near the lowest leaflets; stipules large, linear subulate with a broad base, persistent; petioles ca. 2-3 mm long; leaflets 25-50 pairs, overlapping; lamina ca. 3-8 × 0.8-1.2 mm, subfalcate-oblong, truncate, obliquely acute at base, mucronate at apex, entire, membranous, glabrous, coriaceous, sessile. Inflorescences in axillary racemes, ca. 30-50 cm long; bracts 1 pair, stipulaceous; bracteoles 2, minute; Flowers solitary or in pairs, yellow; sepals lanceolate-ovate, acuminate, finely pilose externally; petals elliptic-orbicular. Pods ca. 1.5-4.5 × 0.25-0.45 cm, strap-shaped, flat, rostrate at apex, covered with fine bristles, dehiscent. Seeds compressed, obliquely oblong.

USES:

It is used to treat a number of ailments, like inflammation, bronchitis, joint pain, diabetes, anaemia, fever, and pneumonia.

Desmodium gangeticum (L.) DC. Fabaceae

Sal Leaved Desmodium, Shalparni, Salpalru, Pulladi

Erect undershrubs, up to 1.2 m tall; stem angled, hairy when young, glabrescent on ageing. Leaves 1-foliolate, 5-10 × 1-5 cm, ovate to elliptic, base rounded, apex acute, glabrescent above, densely appressed hairy below, often blotched with grey; petiole to 2.5 cm long; stipules to 8 mm long; linear-lanceolate. Flowers white, fascicled in terminal and axillary, 15-25 cm long racemes; pedicels 3-4 mm long; bracts subulate. Calyx campanulate; lobes 1.5-2 mm long, triangular, tomentose without. Corolla c. 4 mm across. Stamens diadelphous. Ovary 3-4 mm long, compressed, hairy without; stigma capitate. Pods 1.2-2.3 × 0.1-0.15 cm, moniliform, 4-7-jointed, reticulate, slightly pubescent with hooked hairs, constricted in ventral suture.

USES:

The plant has febrifuge, aphrodisiac, analgesic, diuretic, anti-inflammatory, and haemorrhagic properties. · It is used in postnatal complaints.

Desmodium triflorum (L.) DC. Fabaceae

Lesser Clover-leaf, Kudalia, Sirupulldi

Perennial herb, creeping, branches prostrate, rooting at nodes, pubescent. Leaf trifoliolate, petiole 5-7 mm long, leaflets 5-7 mm long, 4-5 mm broad, obovate, cuneate, truncate or emarginate, glabrous above, pubescent below; stipules 3.5-4.5 mm long, obovate, acuminate. Inflorescence axillary fascicles of 1-5 flowers, bracts ovate, acute, ciliate, pedicel 4.5-5.0 mm long. Calyx 3.5-4.5 mm long, pubescent, teeth longer than the tube. Corolla 5 mm long, pink or white. Fruit 10-16 mm long, 3.5-4.5 mm broad, upper suture straight, lower indented, articles 3-5, as broad as long, puberulous.

USES:

Plant ground in milk is given as a general tonic to ladies. Plant decoction is used for rheumatism and arthritis. Plant extract is recommended to increase the digestive power of children. Whole plant decoction is consumed as a tonic for general weakness.

Diospyros montana Roxb. Ebenaceae

Bombay Ebony, Dasaunda, Karunthuvalisu

Dioecious trees, to 20 m high, bark smooth, grey or yellowish-grey. Leaves simple, alternate, estipulate; petiole 5-10 mm, slender, grooved above, glabrous; lamina 4-10 × 2-4 cm, ovate, ovate-oblong, elliptic-oblong, base round, obtuse, subcordate or acute, apex obtuse, subacute or obtusely acuminate, margin entire, chartaceous, more or less softly pubescent when young, glabrous at maturity; lateral nerves 5-10 pairs, pinnate, slender, faint, intercostae reticulate, faint. Flowers unisexual, white; male flowers: 2-6 in axillary umbels; peduncle to 5 mm; calyx 3 mm long; lobes 4, ovate, imbricate, thick, margin ciliate, obtuse; corolla 6-7 mm long, greenish-yellow, urceolate, glabrous; lobes 4, ovate, imbricate, 2.5 mm long, subacute; stamens 16, roughly in 8 pairs, unequal; anthers lanceolate, awned; pistillode conical, pointed; female flowers: solitary, axillary; pedicel to 4 mm; calyx and corolla as in male flowers; staminodes 4, 5 mm, linear-lanceolate; ovary superior, 0.7 mm across, globose, glabrous, 8-celled, ovule one in each cell; styles 4, 1.5 mm long; stigma capitate; Fruit a berry, 1.5-2 cm across, ovoid to obconical, yellow to orange, glaucous; calyx foliaceous, slightly enlarged, foliaceous, reflexed; seeds 3-6, rough, black.

USES:

Root bark ground with lime juice is given for three days to expel urinary, kidney and bladder stones. Bark cooked with rice or its juice is taken for jaundice. Bark extract with lime juice is consumed for kidney and bladder stone.

Dipteracanthus patulus (Jacq.) Acanthaceae

Perennial herb, mostly decumbent, sometimes shrubby. Stems up to 60 cm long, trailing to erect, almost hairless, velvety or hairy, especially when young, sometimes with stalked glands. Leaves opposite, ovate to more or less narrowly elliptic, 1-6 cm long, sparsely to densely velvety, particularly along the veins, sometimes with scattered capitate glands; petiole 2-32 mm long. Flowers axillary, solitary or 3-5 together. Calyx 3-8 mm long; lobes narrowly triangular. Corolla short-lived, falling in the morning, white, mauve or purple; tube 10-34 mm long, widening at the throat; lobes 4-15 mm, more or less broadly ovate-elliptic; margin entire to crenate-dentate. Capsule clavate, 11-22 mm long, hairless.

USES:

Leaf paste is applied externally for wounds, cuts, poisonous bites and skin diseases.

Diplocyclos palmatus (L.) C. Jeffrey Cucurbitaceae

Shivalingi, Ivirali

Perennial, monoecious herb climbing by bifid tendrils; stem up to 6 m long, young stems spotted with darker green. Leaves alternate, simple; stipules absent; petiole 2-10 cm long; blade broadly ovate, palmately 5(–7)-lobed, up to 14 cm × 15 cm; base cordate; lobes narrowly elliptical or elliptical, margin sinuate-dentate. Inflorescence an axillary cluster, with usually both male and female flowers in same axil. Flowers unisexual, regular, 5-merous, corolla white to greenish-yellow; male flowers pedicellate, with 3 free stamens; female flowers subsessile, with inferior, 1-celled ovary, stigma 3- lobed. Fruit a subglobose, indehiscent berry 1.5-2.5 cm in diameter, solitary or clustered, red with silvery white longitudinal stripes.

USES:

The fruits and leaves are used to cure stomachache; stems are used as an expectorant, fruits are used as a laxative and seeds as febrifuge. The aerial plant parts are potentially utilized for several purposes such as Aphrodisiac, Constipation, Stomach problem, Diarrhoea and Malarial fever.

Dodonaea viscosa (L.) Jacq. Sapindaceae

Australian Native Hop, Sinatha, Viraali

Evergreen shrubs, with gland-dotted branchlets. Leaves alternate, simple; stipules absent; petiole very short, up to 2.5 mm long, or absent; blade oblanceolate or broadly to narrowly elliptical, (1–)4-13 cm × (0.5–)1.5-4 cm, narrowly cuneate at base, obtuse but minutely apiculate at apex, margins entire, both surfaces glabrous but glandular and coated (especially when young) with viscid glandular exudate, with a conspicuous midrib on both sides and 15-20(–30) often indistinct pairs of lateral veins. Inflorescence a loose thyrsoid panicle at the end of twigs. Flowers bisexual or unisexual, whitish to greenish-yellow; pedicel 8-15 mm long; sepals 3-4, free, 2-2.5 mm long; petals absent; stamens 7(–9), filaments very short, anthers oblong, up to 3 mm long in male flowers, up to 2 mm long in bisexual flowers and reduced to staminodes or completely lacking in female flowers; ovary superior, oblong in outline, flattened, 2-3-celled, strongly rudimentary in male flowers, style 2-3-lobed. Fruit a 2-3-winged papery capsule, 15-23 mm × 18-25 mm, white or straw-coloured to brown or purplish, dehiscent by splitting along 2-3 central septa, each cell 2-seeded. Seeds subglobose, more or less compressed, ca. 3 mm in diameter, black.

USES:

The leaves are used as anti-inflammatory, anti-ulcer, anti-bacterial and antifungal agents and in the treatment of fractures Even the green stems and leaves readily catch fire.

Drosera burmannii Vahl Droseraceae

Sundew, Sanadyoo paudha, Azhukunni

Perennial, monoecious herb climbing by bifid tendrils; Stem up to 6 m long, young stems spotted with darker green. Leaves alternate, simple; stipules absent; petiole 2-10 cm long; blade broadly ovate, palmately 5(-7)-lobed, up to 14 cm × 15 cm, base cordate; lobes narrowly elliptical or elliptical, margin sinuate-dentate. Inflorescence an axillary cluster, with usually both male and female flowers in the same axil. Flowers unisexual, regular, 5-merous; corolla white to greenish-yellow. Male flowers pedicellate; with 3 free stamens; female flowers subsessile, with inferior, 1-celled; stigma 3-lobed. Fruit a subglobose, indehiscent berry, 1.5-2.5 cm in diameter, solitary or clustered, red with silvery white longitudinal stripes.

USES:

It is traditionally used to treat respiratory conditions like chronic bronchitis, asthma, and whooping cough. It acts as an antispasmodic and expectorant.

Eclipta prostrata (L.) L. Asteraceae

False Daisy, Bhrngarajah, Vellai Karisalanganni

Erect or prostrate herb, rooting at the nodes. Leaves simple, opposite, ovate to oblong-lanceolate, 2-10 cm long, 1-3 cm wide, apex acute or blunt, base attenuate, margin entire or slightly serrate, pubescent, mostly sessile, the lower leaves sometimes shortpetioled, basally swollen hairs on both surfaces, veins prominent.. Heads heterogamous, up to 1 cm in diameter, a cluster of sessile white flowers, in upper axils or terminal, solitary or two heads together. Peduncle, thickened at the top, variable in length, 0.5-7 cm long, hairy. Involucral bracts 5-6, green, ovate, in two rows, outer ones 4-6 mm long, inner ones usually shorter, prominent, hairy. Ray flowers marginal, pistillate, fertile, corolla white, ligulate, 2-3 mm long. Disk flowers numerous, central, perfect, fertile, corolla whitish, tubular, minute, 1.5-2 mm long. Stamens five, separated filaments, anthers coalesced to form a tube around the style.. Fruit cuneate, 3-angled, sparsely pubescent achene.

USES:

Oil prepared form plant juice is used as hair oil for proper growth and colour of hair, also to prevent hair fall. Oil prepared using this plant along with *Vernonia cinerea in gingelly* oil is applied for hair fall. Plant is used for preparing a tambuli which is useful for gastritis (to normalize digestive system). Plant paste with gingelly oil is applied for elephantiasis. Root extract is consumed to induce vomiting and purgation.

Elaeocarpus gaitrus F. Mull. Elaeocarpaceae

Rudraksha, Uthiratcham, Rudraksham

It is a large evergreen tree up to 25 m, often with buttress roots. Leaves oblong–lanceolate, sub-entire or irregularly crenate, acute or acuminate. Flowers creamy in dense racemes and mostly from axils of fallen leaves, fringed petals. Anthers linear. Drupes more or less spherical bright blue.

USES:

The beads, bark and leaves are having sedative, antihypertensive, and anti-inflammatory properties, particularly aiding in managing stress, anxiety, epilepsy, migraine, and asthma.

Elephantobus scaber L. Asteraceae

Prickly leaved Elephant's Foot, Adhomukha, Yaana chuvadi

A scapigerous perennial herb up to 60 cm tall, perennial, scabrous; rootstock creeping; plants dichotomously branched. Leaves sessile, in radical rosettes; lamina ca. 3-23 × 1.5-5 cm, obovate, oblong-lanceolate, cuneate-attenuate at base, subacute at apex, margin serrate; cauline leaves alternate, few, undersurface pale-brown tomentose. Heads solitary, glomerate, up to ca. 2 cm across, terminal on ca. 2.5-8 cm long peduncles, each subtended by 3-8-cordate, foliaceous, conduplicate, bracts. Involucral bracts 4-seriate; outer subulate; inner lanceolate; Florets pink-purple. Achenes truncate, 10-ribbed, pilose; pappus 1-seriate, white, comprising 4-5-bristles.

USES:

It is used to stimulate diuresis, reduce fever and eliminate bladder stones, as well as to treat nephritis, edema, dampness, chest pain, pneumonia, scabies, arthralgia and leukemia.

Elettaria candamomum (L.) Maton Zingiberaceae

Cardamom, Elachi, Yelakkaai

A tuberous horizontal rhizome, sending up from eight to twenty erect, simple, smooth, green and shining, perennial stems, which rise from 1.80 to 2.25 m in height, and bear alternate elliptical-lanceolate sheathing leaves. Leaves linear-lanceolate, pubescent below.. The flower-stalk proceeds from the base of the stem, and lies upon the ground, with the flowers arranged in a panicle. Flowers white, in lax panicles from the base of leafy stem. Calyx tubular, split on one side, 3-lobed. Corolla 3-lobed; lateral lobes oblong, upper apically concave; labellum obovate; lip white striped with violet. Staminodes short. The fruit is a three-celled capsule, containing many seeds; during drying it is said to lose three- fourths of its weight.

USES:

Seed powder is given with tender coconut water for urine block and to arrest vomiting. Seed decoction is used for stomach ache and febrile conditions. Seed decoction is a tonic, given for gas trouble and to normalize excretions. Its paste is applied externally for leucoderma and elephantiasis.

Embelia ribes Burm.f. Myrsinaceae

Spurious Peeper-com, Baibadang, Vai vilangam

It is a large scandant shrub with long branches, slender, flexible, terete and long intemodes. Leaves are coriacous, 5 × 2-4 cm long, elliptic or elliptic-lanceolate, shortly and obtusely acuminate, entire, glabrous. Flowers are small, greenish-yellow, numerous in lax panicled racemes. Calyx is minute, sepals connate, broadly triangular, ovate and ciliate. Petals are 5 and free and stamens are 5, but shorter than the petals. Fruits are 2.4-4.0 mm in diameter and globular with warty surface, smooth, succulent.

USES:

The Fruit are used in gastrointestinal issues like indigestion and constipation, aids in weight management and Possesses anti-inflammatory and antioxidant effects. It is also used to manage skin disease, reduce infections and promote overall detoxification.

Ensete superbum (Roxb.) Cheesman Musaceae

Cliff banana, Jungli kela., Kalluvazhai

It is monocarpic, non-stoloniferous, unbranched herb. It attains an average height of 2.5 m and is characterized by an enormous swollen base of 2.5-3 m circumference. Leaves short, green petiole with a prominent red midrib. Mature leaves attain an average length and width of 3.5-4 m and 0.5-1.0 m respectively. Petiole short green and red. Leaf sheaths are persistent at the base and closely set scars on the corm. The inflorescence is globose at first (30 cm – 32 cm diameter) later drooping and elongating to one-third the length of the trunk. The bracts are orbicular, dark brown-red, reaching 30 – 32 cm in length, breadth and subtend dense biseriate rows each of 10-15 flowers. The fruits are subcoriaceous 7.5-8.0 cm long, 3.5 cm diameter more or less triangular and contain numerous dark brown seeds. The seeds are sub globose but angled by pressure, 0.8 – 1.3 cm in diameter.

USES:

This plant is used to treat wide range of human diseases like appendicitis, cancer, diabetes, dog bite, kidney stone, hiccups, cardio vascular health, anti- estrogenic and anti-fertility

Erythrina stricta Roxb. Fabaceae

Indian Coral Tree, Paribhadrah, Mullu-Murrukku

Trees, 7-12 m tall. Trunk straight. Branches with short whitish prickles. Leaves pinnately 3-foliolate; stipules deciduous; petiole 12-15 cm, rarely with prickles; terminal leaflet broadly triangular, almost rhomboid, or broadly reniform-oblate, 7-19 × 7-24.5 cm, both surfaces glabrous, lateral veins 5-8 pairs, base nearly cordate, truncate, or broadly cuneate, margin entire, apex acute or caudate with mucro 10-15 mm. Raceme ca. 15 cm, flowers in clusters of 3 or 4. Calyx spathelike, undivided or apex slightly 2-lobed. Corolla red; standard elliptic-lanceolate or ovate-triangular, erect, 3-4.5 cm, subsessile; wings subobovate, 6-8 mm, shorter than calyx; keels much longer than wings. Ovary hairy, stipitate; style narrow, glabrous. Legume 7-12 × 0.7-1.5 cm, glabrous. Seeds 1-3 or more, light or dark brown, reniform.

USES:

Young twig decoction with milk is used for sleeplessness. Bark decoction is given early morning in empty stomach to ladies with menstrual problems. Bark decoction is used in case of threatened abortion. Bark paste is applied on the scalp for hair fall. Also used to treat arthritis, rheumatism and joint pain.

Erythroxylum monogynum Roxb. Erythroxylaceae

Red Cedar, Javadhari, Devadaaru

Small glabrous trees, to 5 m high; bark dark-brown, very rough, lenticellate; branchlets compressed. Leaves simple, alternate, rarely distichous; stipules intrapetiolar, on the arrested shoots imbricating; petioles 3-8 mm, slender, grooved above, glabrous, chartaceous; lamina 2-6.5 × 1-3 cm, obovate or oblanceolate, base attenuate or cuneate, apex obtuse, margin entire, lateral nerves 6-14 pairs, pinnate, slender, faint, intercostae reticulate, faint. Flowers bisexual, white, axillary, solitary or 1-4 in axillary fascicles; sepals 5-6, ovate, acute, glabrous; petals 5-6, white, oblong, with an erect double ligule on inner side, imbricate, stamens 10-12, monadelphous; ovary 3-4-celled, styles 3-4, united into a style with capitate stigmas; ovary superior, 3-4-celled, ovules 1-2 in each cell. Fruit a drupe, some what 3-angled, ellipsoid, apiculate, glabrous, reddish; seed one.

USES:

Heart wood extract is a cooling agent and tonic. Young leaf pounded in oil is applied over the head for biliousness, sleeplessness and vision problems. Heart wood decoction with milk and sugar candy is given for rheumatoid arthritis and polio. Its paste is applied for urticarial and rashes. Oil extracted from heart wood is applied for rheumatism.

Eucalyptus globulus Labill. Myrtaceae

Blue Gum, Nilgir, Karpoora Maram

Evergreen tree 40-70 m tall with straight massive trunk 0.6-2 m in diameter with narrow, irregular crown of large branches and drooping aromatic, camphoraceous foliage. Root system deep and spreading. Bark smoothish, mottled gray, brown, and greenish or bluish, peeling in long strips, at base becoming gray, rough and shaggy, thick, and finely furrowed; inner bark light yellow within thin green layer. Leaves alternate, drooping on flattened yellowish petioles 1.5-4 cm long, narrowly lanceolate, 10-30 cm long, 2.5-5 cm wide, mostly curved, acuminate at tip, acute at base, entire, glabrous, thick, leathery, with fine straight veins and vein inside marlin, shiny dark green on both surfaces. Flowers 1 (rarely 2-3), at leaf base, more than 5 cm across, the very numerous, white stamens ca 12 mm long. Buds top-shaped, 12-15 mm long, 12-25 mm wide. Stamens many, threadlike, white, anthers oblong opening in broad slits with round gland. Pistil with inferior 3-5- celled ovary and long stout style. Capsules single at leaf base, broadly top-shaped or rounded, 1-1.5 cm long, 2-2.5 cm wide, 4-angled, warty. Seeds many, irregularly elliptical, 2-3 mm long, dull black.

USES:

Essential oil extracted from the leaves widely used in cough drops, is antiseptic, rubefacient, and stimulant.

Euphorbia hirta L. Euphorbiaceae

Asthma Herb, Dugdhika, Amman-patcharisi

A creeping to ascending, densely hairy, little-branched annual herb, 15-50 cm tall. Stems semi-erect, several arising from a central tap root, reddish or purplish, with yellow hairs and milky sap. Leaves opposite, 1-4 cm long, 1-1.5 cm wide, oblong lanceolate, with a pointed tip and a finely toothed margin, often reddish, often purple blotched on the upper surface, underside hairy, base rounded and unequal; petiole very short; nerves distinct. Stipules small, linear. Inflorescence dense, globular clusters, 5-10 mm across, of numerous, minute, greenish or pinkish, shortly-stalked flowers without petals arising from the axils of the leaves; peduncles reddish brown. Fruit yellowish, hairy, 3-lobed capsule, 1.25-2 × 1.5 mm, splitting into three I-seeded segments; central column persistent.

USES:

Whole plant decoction with milk is taken for burning sensation in arms, legs, eyes and head. Plant juice is used increase lactation and for digestive disorders. Whole plant ground in milk is recommended for urinary disorders, leucorrhoea, skin diseases and blood in stool. It also acts as a blood purifier.

Euphorbia tirucalli L. Euphorbiaceae

Indian Tree-spurge, Tirukalli

Trees or shrubs, producing abundant milky latex when injured, 2-6 m tall, dioecious, eventually forming a trunk 10-25 cm DBH with rugose, gray or light bark. Stems green, succulent, finely, longitudinally striate. Leaves alternate, present only on new growth; stipules very small, caducous; petiole ± absent; leaf blade oblong-linear, 7-15 × 0.7-1.5 mm, base attenuate, margin entire, apex obtuse. Cyathia clustered at apex of branches, pedunculate, unisexual; involucral leaves minute, membranous, caducous; involucre turbinate, approximately 2 × 1.5 mm, shortly pubescent inside; glands 5, peltate- ovate or subrounded. Male flowers many, exserted from involucre. Female flower: ovary glabrous, exserted from involucre; styles connate below middle; stigma 2-lobed. Capsule 3-lobed, 8 × 8 mm, smooth, sparsely pilose or glabrous. Seeds ovoid-globose, 4 × 4 mm, smooth; caruncle small.

USES:

Decoction prepared from this plant is used as a bath by rheumatic patients. Young shoot tip juice is given in limited dose for stomachache. Plant juice or latex mixed with butter is used as a laxative.

Evolvulus alsinoides (L.) L. Convolvulaceae

English Speed Wheel, Nilapushpi, Vishnukrandhi

Perennial herb, with woody rootstock; stems slender, spreading or trailing, clothed with long spreading hairs. Leaves oblong, elliptic-oblong or lanceolate, 1-2.5 cm long, subsessile, acute to obtuse basally and apically, pubescent. Flowers on peduncles, mostly equaling the leaves. Sepals lanceolate, acute or acuminate, 2-3 mm long, pubescent. Corolla mostly blue,. Corolla blue, funnel-shaped; limb plicate, 3-7 mm broad. Fruit capsular, globose, glabrous. Seeds black, smooth.

USES:

Whole plant decoction with milk is used as a general health tonic. Whole plant decoction is used for fever, dysentery, cough and rheumatism. Oil prepared from plant juice is used as hair oil. Whole plant extraction with milk is given to increase memory power.

Ferula assafetida L. Apiaceae

Asafoetida, Hing, Perungaayam

Plants 2-3 m tall. Root thick tap root. Leaves large, bipinnate, pubescent; segments oblong, entire, obtuse. Inflorescence densely pubescent. Umbels compact. Petals whitish-yellow, persistent. Fruit broadly oblong to suborbicular, c. 1 cm long, 8 mm broad; vittae inconspicuous, numerous, minute; fruit wing nearly as broad as the seed.

USES:

This spice is used as a digestive aid, in food as a condiment and in pickles. It is used in modern herbalism in the treatment of hysteria, some nervous conditions, bronchitis, asthma and whooping cough. It was at one time employed in the treatment of infantile pneumonia and flatulent colic. The gum resin is antispasmodic, carminative, expectorant, laxative, and sedative.

Ficus auriculata Lour. (Moraceae)

Elephant Ear Fig, Anjira, Atthi

Small spreading tree, without aerial roots. Leaves simple, broadly ovate or suborbicular, pubescent beneath, basal veins 6-7. Figs turbinate, peduncled with longitudinal ridges, on shortleafless branches from the trunk, pubescent, purplish or reddish when ripe; basal bracts 3, triangular. Small spreading tree. Trunk usually short, bark grayish dark brown. Leaf blade suborbicular, acuminate at apex, cordate or rounded at base, pubescent beneath, glabrous above, chartaceous, basal nerves 6-7, lateral nerves 4-5, tertiary nerves transverse and parallel. Figs obconic, with longitudinal ridges, pubescent brown, white spotted, basal bracts 3, triangular, orifice sunken and covered by 6 triangular apical bracts.

USES:

Bark decoction and fruit are beneficial for gastric ulcers and burning urination. Fruit juice is given to expel phlegm. Root extract is applied for ring worm and tinea versicolor. Fried fruit paste is applied for gum problems.

Ficus benghalensis L. Moraceae

Banyan Tree, Bahupada, Alamaram

Trees, evergreen, to 30 m. Roots aerial, often descending to ground level and forming pillar-roots Bark of trunks and older branches brown, smooth. Branchlets puberulent, glabrescent in age. Leaves: stipules stout, 1.5-2.5 cm; petiole 1.5-7 cm. Leaf blade ovate, 10-30 × 7-20 cm, leathery, base cordate, margins entire, apex obtuse; surfaces abaxially puberulent, adaxially glabrous; basal veins (2-)3-4 pairs, 1/3-1/2 length of blade, reticulations regular; lateral veins 5-6(-7) pairs. Syconia paired, sessile, orange or red, depressed-globose, 1.5-2 × 2-2.5 mm, pubescent; subtending bracts ovate, 3-7 mm, puberulous; ostiole closed by 3 flat or nearly umbonate apical bracts 3-4 mm wide.

USES:

White terminal portion of prop root ground in milk is given for burning sensation and is a general tonic. Paste prepared form its bark, castor oil, bee wax and turmeric is used as quick healer for cracks in feet. Oil prepared using its latex and milk is used to prevent hair fall and for sleeplessness.

Ficus racemosa L. Moraceae

Country Fig, Sadaphala, Atthi

It is a medium-size deciduous tree, without aerial leaves. Leaves simple, alternate, stipules 12-18 mm long, lanceolate, linear-lanceolate, pubescent, often persistent on young shoots; petiole 10-50 mm long, slender, grooved above, becoming brown scurfy; lamina 6-15 × 3.5-6 cm, ovate, obovate, elliptic-oblong, elliptic-lanceolate, elliptic-ovate or oblong-ovate, base acute, obtuse or cuneate, apex narrowed, blunt or acute, margin entire, membranous, glabrous, blistered appearance on drying; 3- ribbed from base, 4-8 pairs, slender, pinnate, prominent beneath, intercostae reticulate, obscure. Male flowers near the mouth of receptacles, in 2-3 rings, sessile, much compressed; tepals 3-4, dentate-lacerate, lobes jointed below, red, glabrous; stamens 2, exserted; filaments 1 mm, connate below; anthers oblong, parallel; female flowers sessile or very shortly stalked among gall flowers; tepals 3-4, dentate-lacerate, lobes jointed below, red, glabrous, ovary superior, sessile or substipitate, red spotted; style 2-3 mm long, glabrous, simple; stigma clavate; gall flowers long stalked; ovary dark red, rough; style short. Syconium 2.5 × 2 cm, orange, pink or dark crimson; achene granulate

USES:

Bark cooked along with rice is given for biliousness, for cooling effect and also as health tonic. Fruit juice is used for gastritis. Bark paste is applied to ulcer or boils on body due to excessive heat. Water oozing out from the cut root is used for gastric ulcers; it has cooling effect and it is a tonic. Leaf decoction is used to wash ulcers.

Ficus religiosa L. Moraceae

Peepal Tree, Ashvatha, Arasu

Large glabrous tree 15-25 m tall, without aerial roots. Stipules ovate, small, apex acute. Petiole slender, as long as or longer than leaf blade, articulate; leaf blade triangular- ovate, 9-17 × 8-12 cm, leathery, abaxially green, adaxially dark green and shiny, base broadly cuneate to ± cordate, margin entire or undulate, apex acute to caudate with a 2-5 cm cauda; basal lateral veins 2, secondary veins 5-7 on each side of midvein. Figs axillary on leafy branchlets, paired or solitary, red when mature, globose to depressed globose, 1-1.5 cm in diameter, smooth; peduncle 4-9 mm; involucral bracts ovate. Male, gall, and female flowers within same fig. Male flowers: few, near apical pore, sessile; calyx 2- or 3-lobed, margin revolute; stamen 1; filament short. Gall flowers: pedicellate; calyx 3- or 4-lobed; ovary globose, smooth; style short; stigma enlarged, 2-lobed. Female flowers: sessile; calyx 4-lobed, broadly lanceolate; ovary globose, smooth; style thin; stigma narrow.

USES:

Young shoot tip ground and boiled in milk is given for dysentery and amoebiasis. Bark decoction is given for small pox, mouthulcers, rashes, urticarial and release of sperm on dreams. Bark extract has antibacterial and aphrodisiac properties; it is given for diarrhoea and gastroenteritis.

Fluggea leucopyrus Willd. Euphorbiaceae

Spinuous Fluggea, Panduraphali, Vellai poola

A large rigid bush with some what straggling branches; branchlets angular, slender, leafy, usually ending in sharp spines. Leaves 1.5-2.5 × 1-1.5 cm obovate or obcordate or roundate, glaucous and reticulately veined beneath. Flowers in axillaru clusters froma crowd of minute bracts the males very numerous, females usually 1-5. Perianth lobes 5, Stamens usually 5, free, usually with 2-3-fid pistillode. Ovules 2 per locule; styles 3, shortly connate below, once or twice branched, usually spreading. Fruit drupaceous, globose 0.8 cm in diameter, smooth, fleshy, turning white with age.

USES:

Leaf decoction is given to stop dysentery. Ripe fruits eaten raw.

Fumaria indica (Hausskn.) Pugsley Fumaricaceae

Pitparpa

An erect herb ca. 30 cm tall. Radical leaves petiole is 1-7 cm long, decompound, petiolule is 1to15 mm long; lobule linear, oblong, 0.56x0.51 mm, acute, mucronate; both surfaces of leaves are sedge green. Cauline leaves petiole is 0.8-6 cm long, petiolule is 1-13 mm long; lobule linear, oblong, 0.5-6x0.5-4mm, acute, mucronate; both surfaces of leaves are sedge green. Raceme is 0.7-4.5 cm long, 5-23 flowered. Flowers yellow, 4-6 mm long. Sepals 2, caducous, imbricate. Petals 4, 3-4.5x0.5-1 mm, imbricates, upper & lower petal tips are suborbicular while inner are coherent at tips, lower are narrow. Stamens 6, diadelphous, 3 on each side of carpel. Carpel ellipsoid, glabrous, ovary 1-1.5mm long, stigma 0.5-1 mm long, style 1.5-3 mm long. Nut suborbicular, rounded, rugose, 2x2mm; seeds somewhat rounded brown, 1 in number, 1mm in diameter.

USES:

The plant is used to treat purify blood, chronic dysentery and other digestive ailments, urinary disorders and menstrual irregularities, skin diseases, rashes and eczema.

Garcinia gummi-gutta (L.) Robs. Clusiaceae

Malabar Gamborge, Vrikshamlah, Kodampuli

Evergreen trees up to 15 m tall.. Outer bark reddish brown, lenticellate; blaze reddish. Branches drooping; young branchlets subterete, glabrous. Latex yellow, profuse. Leaves simple, opposite, decussate; petiole 5-1.6 cm long, planoconvex or shallowly canaliculate above, slightly sheathing at base; lamina 5-13 × 2-6 cm; variable in shape from narrow elliptic, oblanceolate to obovate, apex usually acute, sometimes obtuse, base cuneate to attenuate, coriaceous or subcoriaceous; secondary nerves not prominent on both sides. Flowers polygamous, in axillary or terminal clusters; calyx cream; petals pink. Male flowers fascicled, pale white or green; sepals 4; petals 4; stamens 12-20, inserted on prominent receptacle. Female flowers 1-3 together; sepals 4, petals 4; staminodes in a ring; ovary 8-11-celled. Berry, globose 8-10 grooved, , to 5 cm in diameter; many seeded.

USES:

Fruit juice is taken for improving liver and digestive functions. It is also used for menstrual irregularities. Fruit juice is collected, boiled and preserved in bottles. A small piece of cloth dipped in and it is applied over tonsils for tonsillitis.

Gardenia resinifera Roth Rubiaceae

Gummy Gardenia, Dikamali, Tikkamalli

Small trees up to 5 m tall, stem glabrous covered with yellow resins. Leaves opposite, obovate, obtuse cuneate-rounded 8-20 × 2.5-7.5 cm. Coriaceous; nerves 20-25 pairs, nerves parallel. Flowers fragrant, axillary, solitary, from the axils of uppermost leaves. Calyx up to 1.8 cm long. Corolla white, tuming yellow when matures up tp 7.5 cm across, tube prominently striate, upto 6 cm long, lobes up to3.7 × 1.4 cm, obovate. Stamens 5, inserted at the throat of the corolla, include danthers sessile, up to 1 cm long. Ovary 1- celled, style stout, up to 5.5 cm long;stigmas clavate. Berries up to 3.2 × 3 cm oblong or ellipsoid, smooth, shortly beaked. Seeds numerous, compressed.

USES:

Gum resin from the buds is applied for chronic septic wounds, ulcers, and to remove maggots from the wound. Gum is used for rheumatism, rhinitis, wounds with foul smell, diarrhoea, worms and toothache.

Gloriosa superba L. Liliaceae

Malabar Glory Lily, Agnimukhi, Kannuvalikkodi

Perennial climber with underground tuber. Leaves alternate or opposite, lanceolate-elliptic, 3-15 × 2.2-3, cordate at base, entire at margin, caudate, apex, ending in tendril, glabrous; petiole sessile. Corymbs or solitary, axillary, up to 10 cm long. Flowers ca. 7-9 cm across, yellow-reddish; pedicel up to 9 cm long, erect. Tepals - 6-lobed, lanceolate-oblonceolate to oblong, acuminate at apex. Stamens 6; filament erect, ca. 4.2 cm long; anther lanceolate-oblong, reflexed. Ovary subglobose; 3-loculed; ovules many; style slender, shortely 3-fid, ca. 4 cm long; stigma 3, subcapitate. Capsules ellipsoid- oblong, sepicidal, up to 5 cm long; seeds many, compressed.

USES:

Purified root paste is used for rheumatoid arthritis and joint pain. It is highly abortive. Root paste is applied over the navel region for quick delivery. Tuber paste is applied externally for furuncles.

Glinus oppositifolia (L.) A. DC Molluginaceae

Grishma Sundaraka, Toora

Prostrate or spreading herb; branches, up to 30 long, minutely-pubescent. Leaves whorl, 3-5, oblonceolate-spathulate, 07-23 × 05-08 mm, cuneate at base, entire-serrulate at margin, shortly apiculate, at apex, glabrous;1-nerved; petiole 02 mm long. Flowers ca. 06 mm across, greenish-white, cyme axillary, 5-6, rarely 7 in fascicled; bracts subulate- lanceolate, 1.2 mm long; peduncel, up to 08 cm long, filiform, glabrous. Tepals 5, elliptic-oblong, 05 mm long. Stamens 5-6, rarely 7; filament flat, 02 mm long; anther H-shaped, 01 mm long. Ovary ellipsoid, 2.6 mm long, glabrous; 3-loculed; ovules many; style 3. distinct, slender, 0.8 mm long; stigma sesile. Capsules ellipsoid-oblong, 05 mm long, glabrous; seeds many, subreniform, granular, bright brown-raddish.

USES:

It has wound healing property and is used for treating joint pains, inflammation, diarrhea and skin diseases and has anti-dermatitic properties. Used to alleviate symptoms of fever and cough.

Gmelina arborea Roxb. Verbenaceae

Gandahar Tree, Gambhari, Kumil

Medium-sized trees about 15 m tall; bark grayish brown; branchlets, petioles, and inflorescences densely yellow-brown tomentose. Branchlets slightly 4-angled when young, becoming terete, lenticellate, leaf scars prominent. Petiole terete, 3.5-10 cm; leaf blade broadly ovate, 8-19 × 4.5-15 cm, thickly papery, base broadly cuneate to subcordate, apex acuminate; veins 3-5 pairs, abaxially prominent. Inflorescences terminal, narrow thyrses; peduncle 15-30 cm. Calyx 3-5 mm, with several black discoid gland patches; teeth 5, sharply triangular. Corolla yellow, 3-4 cm, 2-lipped, sparsely glandular; lower lip 3- lobed, outside yellowish brown puberulent, inside glabrous; upper lip entire or slightly 2- cleft. Ovary glabrous, glandular. Stigma unequally 2-cleft. Drupes yellow when ripe and black when dry, ellipsoid to obovoid.

USES:

Bark cooked with rice is eaten for biliousness and giddiness. Root bark paste is applied for scorpion bite, while decoction for joint pain and various pains. Root powder paste is applied for rheumatism. It is a key ingredient in Dashamoola a formula for inflammatory conditions.

Gynandropsis pentaphylla (L.) DC.(*Cleome gynandra* L.) Capparaceae

Wild spider flower, Ajagandha, Thaivelai

Erect, foetid, glandular-pubescent herb, up to 50 high. Leaves alternate, 3-5-foliolate; leaflets obovate-elliptic to oblong, 2-4 × 1.2-2 cm, cuneate-retuse at base, denticulate at margin, acute-obtuse at apex, minutely glandular-pubescent. Raceme or corymbs terminal, up to 30 cm long. Flowers ca. 1 cm across, white; pedicel, slender, 2.2 cm long; bract 3-foliate, obtuse-oblong. Sepals 4, lanceolate, 6.2 mm long, minutely short glandular-hairy. Petals 4, obovate-stathulate, 1.9 cm long, cuneate at base. Stamens 6; filament filiform, 2.8 cm long, purpule; anther linear-oblong, basifixed, 2.5 mm long. Gynophore, up to 2 cm long. Ovary stalked, subflattened, 1.2 cm long, greenish-purple; 1-loculed; ovules many; style ca. 02 mm long; stigma capitate-peltate. 01 mm across. Capsules linear-cylindric, up to 6 cm long, short-glandular, 2-valved; seeds many, circular, 1.1 mm long, concentric stirate.

USES:

Seed decoction is recommended for fever and biliousness. 1-2 drops of leaf juice mixed with salt is poured into ear for earache. Whole plant boiled in coconut oil is applied for bruises, septic wounds, ulcers and leprosy.

Gymnema sylvestre R. Br ex Schult. Asclepiadaceae

Indian Ipecacuanha, Gurmar, Siru-kurinjan

It is a climber, much branched. Stems to 3 m; young branchlets pubescent, glabrescent. Petiole 3-12 mm; leaf blade obovate to ovate, 3-8.5 × 1.5-5.5 cm, thick papery, adaxially pubescent to glabrous except for groove of midvein, abaxially tomentose to glabrous except for veins; lateral veins 4 or 5 pairs, convergent. Cymes much shorter than leaves, pubescent; peduncle 2-5 mm, rachis with close-spaced spiral of pedicel scars. Sepals ovate, ciliate. Corolla greenish white; lobes ovate, glabrous; appendages exserted. Stigma head exserted. Follicles mostly solitary, broadly lanceolate in outline, 5-9 × ca. 2 cm, glabrous, beak acuminate. Seeds ovate, ca. 8 × 4 mm; coma silky white, ca. 3.5 cm.

USES:

The plant has anti-obesity and anti-diabetic properties and to act as laxative and cough suppressant.

Hedyotis corymbosa (L.) Lam. Rubiaceae

Diamond flower, Daman pappar, Kattu Chayaver

Erect annual herbs upto 30 cm. Leaves opposite, sessile; stipules small, membranous, truncate, bristles few; lamina tapering at base, linear, linear-lanceolate, surface glabrous adaxially, scabrescent abaxially, apex acute, margins recurved, scabrous. Flowers: 0.2 cm long, axillary, 2-3 fascicled, often solitary; peduncles 0.5-1.2 cm long, pedicels 0.2-0.5 cm long; bracts at the base of pedicels 0.1 cm long, filiform; calyxsparsely pubescent, 0.2 cm long; tube ovoid, 0.1 cm long; lobes 4, 0.1 cm long, triangular, acute, ciliate; corolla- tubular, 0.2 cm long, tube slender, lobes 4, oblong, 0.05 cm long, obtuse; stamens- 4, 0.4 cm long, filament short, included; style- 0.1 cm long; stigma bifid, hairy. Capsules globose, didymous, 0.2 cm across, top not protruded beyond calyx, glabrous; seeds, pale-brown, 0.01 cm long.

USES:

The plant is known to clear heat and toxins, activate blood circulation, promote diuresis and relieve stranguria. In Chinese medicine, it is used to treat viral infections, cancer problems and skin ailments. The juice of the plant is applied to the hands and feet to cool them, when the patient has a fever.

Helicteres isora L. Sterculiaceae

East Indian Screw tree, Marorphali, Edampiri-valampiri

Shrubs or small trees, up to 7.5 m tall. Leaf blade broadly oblong-obovate, or almost circular, 10-23 × 7-17 cm, thinly leathery, base rounded or obliquely cordate, margin serrate, apex truncate with shortly acuminate tip. Inflorescences axillary. Flowers 3.5-4 cm long. Epicalyx lobes subulate, to 7 mm. Sepals 17-20 mm, usually 4-5-lobed, lobes triangular, 2-lipped, densely stellate hairy. Petals red or purple, unequal in length, upper 2 larger, 1.2-1.5 cm, obliquely falcate. Androgynophore 3-4 cm, curved at tip. Stamens 10; staminodes 5, equal to filaments in length. Ovary slightly verrucose, spirally twisted after pollination. Capsule on ca. 2.5 cm stalk, cylindric, conspicuously spirally twisted, black when mature, 4-8 × 0.7-1 cm, apex acute, with beak to 1 cm, densely stellate puberulent at first, glabrescent.

USES:

Tender twig boiled with coconut oil is used for paralysis of limbs and arms. Root bark decoction is given to women after delivery for a day, in order to contract the uterus. Tender shoot tip juice is given to small children as one of the sour agents.

Heliotropium indicum L. Boraginaceae

Indian Heliotrope, Hastishundi, Thel kodukku

Erect, branched, pubescent herbs, Leaves to 11x 6cm, sub-deltoid or broadly ovate, base truncate, margin undulate, apex acute, chartaceous; petiole 3l-6 cm long. Spikes terminal or leaf- opposed, curved, to 20cm long. Flowers sessile. Calyx-lobes 5, , 1.5-2 mm long, linear, acute. Corolla pale violet or pink, salver-form; tube c. 1m long, ovate. Stamens 5, subsessile, free. Ovary weakly 4-lobed; stigma conical. Nutlets connate in pairs, separating later, ribbed, each nutlet 2-3 mm long.

USES:

It helps in curing fever, insect bites, stings, diarrhea, skin rashes, menstrual issues etc.

Hemidesmus indicus (L.) R. Br. Periplocaceae

Indian Sarasapalli, Magrabu, Nannaari

Perennial twining undershrub, with woody rootstock. Branches terete, glabrous or pubescent. Leaves simple, highly variable, linear-lanceolate to elliptic-oblong, , 5-10 × 0.8-1.8 cm frequently variegated with white along the veins. Flowers small, in axillary cymes. Calyx 5-lobed; lobes broadly ovate. Corolla 5-6.5 mm long, rotate, purplish-brown within; lobes 5. Corona of five fleshly, broadly truncate scales. Filaments free; anthers connivent. Follicle 10 × . 0.6 cm, tapering to a point. Seeds 6-8.5 mm long, ovate, oblong, flattened, black, coma white, 2.5 cm long.

USES:

Inner portion of root ground with milk is used as a health tonic, for urinary diseases, urinary stones, burning during urination, weakness, gastritis and as a blood purifier. Root decoction with milk is a good tonic for children and pregnant women. Root oil is used to message the body of small children. Inner root paste is applied for herpes and snake bite. Fried root powder is taken for thrush in tongue of children.

Hemigraphis alternata (Burm.f.) T. Anderson Acanthaceae

Murikootti

Ascending or creeping herb, the stems pilosulous, rooting at the lower nodes. Leaves on petioles 1-5 cm long, blades mostly 3-6 cm long, 2-5 cm wide, broadly ovate to ovate-oblong, acute or obtuse, cordate at base, densely puberulent, sparingly pilosulous or glabrate, the margins crenate. Inflorescences are axillary spikes on a peduncle to 1.5 cm long and 1 cm wide. Flowers generally two per bract. Calyx segments slightly unequal the, linear-subulate, up to 9 mm long, peeled, ciliate. Corolla white, about 1.5 cm long, tubular ventricose; lobes 5, subequal. Stamens 4, didynamous. Capsule slender; seeds 4-20.

USES:

Plant or leaf paste is a wound healer and is applied to arrest bleeding from wounds. The plant can lower blood glucose levels. It is used in treating anemia, hemorrids and as a diuretic.

Hibiscus rosa-sinensis L. Malvaceae

Chinese Hibiscus, Raktapushpi, Semparutthi

A glabrate shrub 3 m tall. Leaves 4-8 cm long, 2-5 cm broad, lobed or not, ovate to broadly ovate, serrate-dentate; stipules 5-10 mm long, linear; petiole 0.5-2 cm long. Flowers axillary, solitary, erect or subpendulous, single or double; pedicel 1-8 cm long, articulate near the top. Epicalyx segments 6-9, 5-15 mm long, linear or linear-lanceolate. Calyx tubular, campanulate, 1.5-2.5 cm long; lobes 5-15 mm long, triangular to lanceolate. Corolla 4-9 cm across, red, with or without dark centre, petals obovate or oblong-obovate, 5-9 cm long, 3-7 cm broad, obtuse or irregularly lobed at apex. Staminal column exserted, antherifierous near the apex. Fruits rarely formed.

USES:

Leaf paste with rice seed is applied over the warts and carbuncles for pus release, quick heal, also for ringworm. Leaf paste with lime juice is applied for carbuncles. Leaf mucilage is a laxative, used for constipation and as liver medicine. Shade dried flower powder is used with milk for piles and with honey for heavy menstrual bleeding.

Holarrhena pubescens (Buch.-Ham.) Wall.exG. Don Apocynaceae

Ivory Tree, Kutja, Veppaalai

Medium sized trees up to 15 m tall, with abundant white latex in all parts. Leaves opposite, simple, ovate or elliptical, 1.5-20 cm × 1.5-11 cm, base cuneate to rounded, apex acuminate to acute, shortly hairy to glabrous, pinnately veined with 5-25 pairs of lateral veins. Inflorescence a axillary cyme, Flowers 5-merous, fragrant. Sepals elliptical to linear, 2-12 mm long, free or fused at base, shortly hairy Corolla-tube cylindrical, 9-19 mm × 0.5-2 mm, shortly hairy outside and inside from the throat to the insertion of the stamens, lobes elliptical, 10--24mm × 3-8 mm, white. Stamens inserted 1.5-2.5 mm from the corolla base, included. Ovary with 2 carpels connate at extreme base, style cylindrical, ca. 1 mm long, pistil head ovoid, stigma 2-lobed. Fruit composed of 2 long and slender follicles 20--38 cm × 2-9 mm, pendulous, dehiscent, pale grey to dark brown, manyseeded.

USES:

It is used for the treatment of amoebic dysentery, diarrhoea, irritable bowel syndrome, bleeding piles, and used for respiratory issues like asthama and bronchitis.

Holoptelea integriflia (Roxb.) Planch. Ulmaceae

Indian Elm., Putikaranjah, Aya

Large deciduous trees, to 25 m high, Leaves simple, pubescent; lamina 6-12.5 × 2.5-6.5 cm, ovate-oblong, ovate or elliptic-ovate, base rounded or subcordate, apex acuminate, margin entire, lateral nerves 6-9 pairs. Flowers polygamous, appear before leaves, 5-8 mm across, greenish-purple, in axillary fascicles. Tepals 4 or 5, free, anthers pubescent; female flowers with longer pedicels. Ovary superior, compressed, long stipitate, 2-winged, 1-celled, ovule 1; style 2 fid. Fruit a samara, 3 cm across, orbicular, wings nerved, glabrous, seed one.

USES:

Oil prepared from its bark is used for chronic ulcers. Bark paste is applied to the pit viper bite and also to arrest bleeding from wounds. Bark decoction is taken in small dose's for tumors, septicemia and liver disorder. Leaf paste is applied for skin diseases and septic wounds.

Homonia riparia Lour. Euphorbiaceae

Willow-Leaved Water Croton, Sherni, Kattalari

It is a shrub growing to 2m tall. The leaves are linear-lance shaped, 12-20 cm long, and 1.5-2 cm wide. Upper surface of the leaves is green and shining, and the lower surface of the leaves is green and shining, and the lower surface brown and hairy. Reddish flowers are born in spikes 5-10cm long, with obovate bracts, 1.5-2 mm long. Male flowers have 0.2 mm long stalks, 3 velvety sepals, 3-4 mm long. Female flowers have 5 oblong sepals. With tapering tips, about 1-2 mm long. The capsules are about 8 mm in diameter, hairy, and borne on solitary, hairy spikes, 5-12 cm long, in leaf axils.

USES:

The plant is used in treating constipation, emesis, piles, bladder stones, gonorrhea, syphilis, and wounds caused by scorpion and fish bites. Pounded leaves are applied as poultice for skin diseases.

Hybanthus enneaspermus (L.) F. V. Muell. Violaceae

Pink ladies slipper, Ratan Purush, Oorithazh Thamarai

Small, perennial herb, with spreading branches. Leaves simple alternate, linear- lanceolate to obovate, 08-13 × 05-06 mm, cuneate at apex, serrulate at margin, acute- subobtuse at apex, glabrous; 6-pairs; petiole sessile. Flowers ca. 1.2 across, pink, solitary, axillary; pedicel 08 mm long. Calyx-lobe 5, subequal; lobe lanceolate, hairy. Petals 5; upper elliptic-lanceolate, 3.2 mm long, acuminate at apex; laterals oblong, 05 mm long, subobtuse at apex; lower obovate-suborbicular, 1.5 cm long. Stamens 5; filament 1.2 mm long; anther free, gibbose, 1.5 mm long. Ovary ovoid, 3-angled, 01 mm across; 3-loculed; ovules 2-4 per cell; style suberect, 1.5 mm long; stigma subglobose, 0.5 mm long. Capsules globose, 0.5 mm across, glabrescent; seeds 9-13, ellipsoid-obovoid, many ribbed.

USES:

Whole plant ground in milk is given for 48 days to male suffering from sexual weakness and women for leucorrhoea. It is also given to increase sperm count and as a general tonic. Whole plant ground with milk and sugar is used for sexual vigour and urinary tract inspection. Oil prepared from plant extract is good for hair growth and colour.

Hydnocarpus pendandra (Buch.-Ham.) Oken. Flacourtiaceae

Jangli Almond, Chaulmoogras, Niradi/ Maravetti

Dioecious trees, to 25 m high, bole fluted; bark 4-5 mm thick, surface pale brown mottled with white patches, smooth, brittle; blaze creamy yellow; branchlets pubescent. Leaves simple, alternate; stipules small, lateral, cauducous; petiole 8-10 mm, stout, pubescent, grooved above; lamina 8-17 × 3-8 cm, elliptic, elliptic-lanceolate or elliptic- oblong, base acute, obtuse or round, apex acuminate, margin crenate, serrate or entire, glabrous, coriaceous, lateral nerves 6-8 pairs, pinnate, prominent, intercostae scalariform, prominent. Flowers unisexual, 5-10 mm across, greenish-yellow, solitary or in small fascicle; sepals 5, broadly ovate, pubescent; petals 5, broadly ovate, greenish-yellow, fringed with soft white hairs, with a scale at the base; scales half as long as petals, densely hairy; stamens 5-15; filaments subulate, hairy at the base; anthers of female flowers possess no pollen; ovary globose, 1-celled, tomentose; stigma sessile and two lobed. Fruit a berry 5-7 cm across, globose, rind thick, brown, rough with uneven surface; seeds numerous, yellowish, obtusely angular.

USES:

It is primarily used for treating leprosy and other skin disorders and is also used to treat other indications including constipation, inflammation, blood disorders, and worm infestations.

Hygrophila schulli (Buch.-Ham.)M. R. Almeida & S. M. Almeida Acanthaceae

Long leaved Barleria, Tal-makhana, Neermulli

It is an aquatic perennial herb to 60 cm high. stem sparsely branched, 4-angled with 6 sharp spine at nodes, hispid. Leaves opposite, linear-lanceolate to oblonceolate, 4-11.5 × 03-7 cm, cuneate, attunate at base, serrulate, undulate at margin, acute-acuminate at apex. Flowers ca. 2 cm across, blue-violet axillary, cluster, sessile; bracts lanceolate, leafly, 1.7 cm long, ; bracteole linear, 4 mm, margin long ciliate. Calyx 4-lobe; outer 2, lanceolate, 1.3 cm, upper densely hirsute; inner 2, linear-lanceolate, 1.5 mm. Corolla 5, deeply 2- lipped; tube 1 cm lonh; upper 2; lower 3, ventricise, wooly-tomentose. Stamens 4; filament 05-08 mm; anther oblong-obovate, 2 mm, upper surface yellowish-purple; bluish, hispid- tomentose, beneath. Ovary elliptic-obovate, 3 mm, glabrous; 2-loculed; ovules 2; style cylindric, 1.8 cm long, pubescent; stigms curved at tip, pink. Capsules cylindric-slender, ca. 09 cm long; seeds many, orbicular.

USES:

The decoction of root is given as a diuretic for kidney stones. One spoonful of seed extract is given for 15 days will cure sexual weakness among males. The plant has antiglycemic and antitumour properties.

Hyptis suaveolens (L.) Poir. Lamiaceae

Ganga basil, Bhustrna, Kattuthulasi

It is an erect plant up to1.5 m tall, Leaves 2-10 cm long and 1-7 cm wide ovate, elliptic slightly cordate, oppositely arranged and borne on petioles that are 5-40 mm long, margins are shallowly toothed and pubescent. Flowers are small (5-7 mm long), pinkish, bluish-purple or lavender in colour and arranged at the axils in 1-5 flowered clusters. Pedicels are 1-5 mm long. Flowers are tubular in appearance with two 'lips'. The upper lip is divided into two lobes and the lower lip divided into three lobes.. It has five lobes, each being topped with a short bristle about 5 mm long and turns brown in colour. The fruit is a schizocarp that divides into two 'seeds'. These 'seeds' 3-4 mm long and 2.5-3 mm in wide are dark brown to black in colour with whitish markings at one end.

USES:

Leaf paste is applied for rashes, to heal fresh wounds and skin diseases. Leaf extract is a liver stimulant. Plant decoction is used for dysentery and diarrhea.

Indigofera tinctoria L. Fabaceae

Indian Indigo, Neeli, Avuri

An erect Shrub, 60-120 cm tall, copiously branched, pubescent, hairs appressed. Leaves 1.2-12.5 cm long; leaflets 7-13, c. 5-26 mm long, elliptic or oblong, glabrous to subglabrous above, appressly pubescent below; petiolule 1 mm long; stipules small. Inflorescence a subsessile raceme, ca. 5-12 cm long. Calyx ca. 1-1.5 mm long, pubescent, teeth as long as the tube. Corolla pink. Vexillum ca. 4-4.5 mm long. Fruit 1.8-3.8 cm long, ca.2.5 mm broad, Fruits cylindric pods, pale greenish grey when young and dark brown straight or slightly curved, glabrous, 10-12-seeded.

USES:

It is used to treatment of fever, liver and spleen disorders, heart palpitations. Used as antidote for snake-bite, scorpion and insect bites. Hair oil is prepared from the stem bark.

Ipomoea obscura (L.) Ker-Gawl. Convolvulaceae

Common Creamy White Bindweed, Dhudhiakalmi, Kuranthalli

Perennial herb with trailing or twining stems up to 3 m long, Leaves distinctly cordate (heart-shaped), 2.5-8.5 cm long, more or less pubescent on both surfaces; apex pointed; margin entire but wavy; petiole op to 11 cm long. Flowers axillary, 1 or 2 together. pale apricot, yellow, cream or white, rarely with a purple centre. Sepals 5, ovate. Corolla funnel-shaped, white to yellow, with a red-purple center. Stamens 5. Capsule globose, tipped with the persistent style base, with ovoid, dark-brown, appressed pubescent seeds.

USES:

Whole plant extract in milk is used as general health tonic. Paste made of whole plant, *Indigofera tinctoria* leaf and sandalwood in rice washed water is applied for swelling due to urticarial and rashes. Plant paste is applied externally for urticarial, rashes and other skin diseases.

Ixora coccinea L. Rubiaceae

Jungle Flame Ixora, Rangan, Vetchi

Large glabrous shrub 3m highwith few branches. Leaves simple, opposite decussate oblong or elliptic, sessile, main nerves 8-12 pairs, coriaceous olive green when dry; stipules cuspidate. Flowers bright scarlet, in dense peduncled, corymbose cymes. Calyx 5- lopped; lobes triangular. Corolla salver-shaped; tube slender, lobes 5, ovate. Stamens 5. Fruits: globose, fleshy, smooth, crowned with the persistent calyx, purple when ripe. Seeds: concave on the ventral surface.

USES:

Flowers bud and cumin seeds are chewed for mouth sores. Root decoction is recommended for washing ulcers and septic wounds. Whole plant decoction with milk is used in case of penis ulcers. Flower juice is poured into eye during eye infection. Root paste is applied for septic blisters.

Jatropha gossypifolia L. Euphorbiaceae

Belly Ache Bush, Dravanti, Adalai

Small shrub, with reddish branchlets. Leaves plamately, 3-5-lobed, young leaves margins; petioles and stipules with glandular-tipped hairs. Leaf petioles are 2-7 cm long and the leaf blades are palmately 3-5 lobed, generally 45-90 × 50-130 mm in size, elliptic, glabrous is denticulate with venations ending in stipitate glandular hairs. Flowers in corymbose cymes. Flowers are radially symmetrical and loaded with nectaries. The calyx five united sepals, ovate to lanceolate, acuminate with stalked capitate glands on the margins, distinctly imbricate and 4 mm long. The corolla five petals obovate and 3-5 mm long. Sepals touch at the base to form a short green tube and have stipitate marginal glands. The fruit trilocular, with a single seed per locule and axile placentation. It is globose, pedicellate, generally bright green and woody at maturity, turning pale green or tan when ripe.

USES:

Seed cause drastic purgation and mental imbalance. Latex is applied for carbuncles, eczema and itches. Young shoot tip cooked or boiled in milk is given for liver, spleen and eye diseases. Paste of fried and powdered seed is applied for ankle and knee pain.

Justicia adhatoda L. Acanthaceae

Malabar Nut, Arasu, Aadaatoda

An evergreen shrub, up to2.5 m.with quadrangular stem. Leaves with 1.5-4.5 cm long petioles; lamina elliptic-lanceolate, 10-20 × 3.5-8 cm, glabrous above, pubescent on nerves beneath, basally attenuate, entire, acute. Flowers white, c. 3 cm long, nearly sessile, in terminal and axillary spikes. Calyx 5-lobed, lobes linear-lanceolate, 6-10 × 2 mm, acute, puberulous, imbricate. Corolla pale-white, tube 1.2-1.5 cm long, pubescent outside, throat villous, limb 2-lipped, upper lip erect, shortly bifid, galeate, lower lip with 3 elliptic, obtuse lobes. Stamens 2, filaments 1-1.5 cm long, anthers oblong, basally apiculate. Ovary oblong, 3 mm long, style 2-2.5 cm long. Capsule stipitate, broadly clavate, 2.5 cm long, 4-seeded.

USES:

Leaf juice mixed with honey is given for cough and phlegm. Flower juice is poured into eye in case of eye infection. Leaf decoction is given for rheumatism, tuberculosis and blood discharge during vomiting or blood in stool. Root, leaf or flower decoction is useful for intermittent fever and lung diseases.

Justicia gendarussa Burm. f. Acanthaceae

Water Willow, Vataghna, Karunochchi

It is an erect much branched undershrub, 0.75 m height with subterete branches. Leaves are simple, lanceolate or linear – lanceolate, 7.5 to 12.5 cm long, glabrous, short petioled, pale green beneath and dark violetgreen above, 8 pairs of main nerves, mid rib and main nerves prominent on the under surface. Stems and branches are dark violet. Flowers in terminal spikes, often forming panicle 5-12.5 cm long from the uppermost leaf axils; white coloured, spotted with purple and clustered in the interrupted spikes. Bracts narrow. Calyx 5- partite3.8-5mm; lobes liner-lanceolate. Corolla white with purple spots, 2-lipped. Stamens 2. Fruits glabrous capsules.

USES:

Whole plant decoction is given internally followed by an external bath for rheumatism, joint pain, body pain and fever. Leaf along with salt and coconut gratings are heated and pressed in the form of pouch for rheumatism. Tender shoot tip juice is given for chronic rheumatism. Leaf decoction is recommended for fever. Leaf paste with salt is heated and pressed for osteoarthritis. The leaves and roots are acrid, febrifuge, thermogenic, emetic, anodyne, emmenagogue, diaphoretic, insecticidal and antipyretic.

Kalanchoe pinnata (Lam.) Pers. Crassulaceae

Air Plant, Parnabija, Ranakalli

An evergreen herb, up to 50 cm with quadrangular stem. Leaves with 1.5-4.5 cm long petioles; lamina elliptic-lanceolate, 10-20 × 3.5-8 cm, glabrous above, pubescent on nerves beneath, basally attenuate, entire, acuinate. Flowers white, c. 3 cm long, nearly sessile, in terminal and axillary spikes. Calyx 5-lobed, lobes linear-lanceolate, 6-10 × c. 2 mm, acute, puberulous, imbricate. Corolla pale-white, tube 1.2-1.5 cm long, pubescent outside, throat villous, limb 2-lipped, upper lip erect, shortly bifid, galeate, lower lip with 3 elliptic, obtuse lobes. Stamens 2, filaments 1-1.5 cm long, anthers oblong, basally apiculate. Ovary oblong, c. 3 mm long, style 2-2.5 cm long. Capsule stipitate, broadly clavate, c. 2.5 cm long, 4-seeded.

USES:

Leaf juice mixed with honey is given for cough and phlegm. Flower juice is poured into eye in case of eye infection. Leaf decoction is given for rheumatism, tuberculosis and blood discharge during vomiting or blood in stool. Root, leaf or flower decoction is useful for intermittent fever and lung diseases.

Kedrostis foetidissima (Jacq.) Cogn., Cucurbitaceae

Monkey Pepper, Appakk:ovai.

Climbers. monoecious; branchlets sparsely pilose Leaves petioled, shortly pubescent or scabrulous, ovate, 2-4 cm long and as broad; base deeply emarginate, apex acute, margin entire, undulate or denticulate, tendril filiform. Male inflorescence usually 2- flowered, 0.5-1.5 cm long. Calyx- tube hemispherical, shortly and sparsely villous, c. 2 mm long, teeth narrowly triangular. Petals yellow, oblong lanceolate, acute, c. 4 mm long. Female inflorescence 1-7 mm long. Fruit deep red when mature, biloculate, emitting foetid smell when ripe, ovoid, sessile, or subsessile 10-25 × 8-12 mm, 1-5 seeded.

USES:

The plant is used for treating ailments like common cold, diarrhoea, asthama and bronchitis.

Lantana camara L. Verbenaceae

Rai Muniya, Unni

Evergreen shrub with rambling or straggling branches, 1-2(-4) m, tall; branches usually minutely or inconspicuously pubescent, unarmed to conspicuously prickly with hooked spines. Leaves opposite, decussate, ovate to ovate-oblong, (2-) 5-10 (-12) cm long, (1.5-) 2-5 cm broad, crenate-serrate, acute to shortly acuminate, ± rugose, scabrid; petiole 5-10 (-12) mm long. Flowering heads axillary, peduncled, umbellate in flower, shorter to exceeding the subtending leaves, 2-3 cm across. Bracts lanceolate to linear, 5-7 mm broad, acute to subulate, rarely a few larger ones also present. Flowers 5-8 mm across, mostly orange or yellow, turning to red or scarlet later. Calyx 2-3 mm long, thin, pubescent. Corolla-tube 7-10 (-12) mm long, pubescent, slightly enlarged and curved above the middle; limb 4 lobed with spreading, ± rounded lobes. Drupe 3-5 mm in diameter, globose; fleshy, black, shining, 2-seeded.

USES:

Leaf juice is used to stop bleeding and to heal cuts, wounds and ulcers. Plant decoction or leaf paste is used for rheumatism. Ground leaf heated with coconut oil is applied for black marks and ulcer caused by bum. Leaf ground with turmeric powder is applied for septic ulcers with foul smell. Leaf cooked with barley is given for ladies after delivery.

Lawsonia inermis L. Lythraceae

Henna, Mehandi, Maruthondri

Fragrant shrub, with tetragonous branches up to 2.5 m tall. Leaves simple, elliptic, ovate or obovate, acute, obtuse or subacuminate, 0. 8-4.4 cm long 0. 2-2.0 cm broad. Flowers small, white or rose, in axillary and terminal panicles 3-22 cm long. Pedicels 2-3.5 mm long. Sepals ovate. Petals 3-4 mm long, 4-5 mm broad, wrinkled. Stamens 8, filaments 4 mm long. Fruit globose capsule 0.5-1 cm in diameter, tipped with persistent style, wall veined.

USES:

Leaf paste is applied on head for dandruff and skin diseases. Oil prepared from leaf juice is applied on head for proper growth of hair and to prevent hair fall. Leaves ground in milk or rice washed water is given for four days from the first day of menses as a blood purifier and for conception. Leaf or stem decoction is given for jaundice and blood in stool.

Leptadenia reticulata (Retz.) Wight & Arn. Asclepiadaceae,

Cork Swallow wort, Jivanti, Palakkodi

Twining shrubs, stems appressed puberlous. Leaves 4-12 × 1, 5-4.5 cm, ovate.ovate-lanceolate acute, entire. rounded, subcordate to truncate at base, subcoriaceous.petioles up to 2.5 cm long. Flowers small greenish, in many-flowered lateral umbellate cymes. Corolla lobes about 0.3 cm long, pale greenish-white. Follicles usually solitary up to 9.5 × 2.5 cm subwoody.ovate-oblong, tape-ring upwards, beak curved.

USES:

Leaves edible. The whole plant decoction is given for paralysis and stroke. Dried tuber decoction is used as a health tonic due to the presence of various phytochemicals.

Limonia acidissima L. *Feronia elephantum* Correa Rutaceae

Wood Apple, Kaithbu, Vilam

Armed deciduous trees, up to 10 m tall.. Leaves imparipinnate, alternate, 1-3 in a cluster, estipulate; rachis 60-80 mm long, stout, glabrous, often narrowly winged; leaflets 4-7, opposite, sessile, estipellate; lamina 1.3-3.8 × 1.3 cm, obovate, base cuneate or acute, apex obtuse, margin entire, glabrous, pellucid-punctate, coriaceous. Flowers polygamous, dull red, 1.3 cm across, in axillary cymes. Calyx small, flat, 5-toothed, pubescent with out, deciduous. Petals 5, free, spreading. Stamens 10-12, inserted round the disc; filaments dilated below, villous on face and margins; anthers linear-oblong; disc thick, annular, pubescent; pistillode short. Ovary superior, oblong, 5-6-celled, at length 1-celled, ovules many; stigma oblong, fusiform. Fruit a berry, 5-7.6 cm across, globose, whitish-brown, rind hard and woody; seeds many.

USES:

It is used mainly as a liver tonic to stimulate the digestive system. The fruit is also astringent, especially when unripe, and a cardiac tonic. The fruit is also seen as an effective treatment for hiccough, sore throat and diseases of the gums. Both the fruit pulp and the powdered rind can be poulticed onto bites and stings of venomous insects.

Lobelia nicotianef olia L. Nala,

Kattu Puhaielai

A shrub, stem stout, hollow, 3-5 meter high, occasionally branched in the upper part. Leaves very large, up to 45 cm long in lower part of stem, smaller upwards, margins not entire; the main nerve of the leaves whitish. Flowers large white, in very large terminal bunches. Fruit 8 mm, roundish. Seeds many, small, yellowish, and brown.

USES:

It is used in treating epilepsy, a recent study have found that its antiepileptic activity is due to Lobeline. It stimulates respiration and is used for reviving respiration in cases of failures of respiration caused by excessive narcotics, anesthesia, gases, etc.

Madhuca longifolia (J. Koenig ex L.) J. F. Macbr.var. latifolia(Roxb.) A. Chev

Mahua, Kattu Eluppai

Tall trees, to 20 m high. Leaves simple, alternate, clustered at the end of branchlets;stipules lateral, lanceolate, cauducous; petiole 20-40 mm long, stout, glabrous lamina 10-25 × 6-12 cm, elliptic, base round, obtuse or cuneate, apex acute, acuminate or apiculate, margin entire, glabrous, coriaceous; nerves 9-14 pairs, parallel, prominent. Flowers creamy, axillary. Calyx- lobes 4, obovate, fulvous tomentose, acute. Corolla- tube 8 mm long, lobes 6-12, oblong. Stamens 16, roughly in 3 series. anthers 3.5 mm, sparsely hairy; connectives apiculate. Ovary superior, 3 mm long, 6-celled; style 2.5 cm long, exserted, stigma obtuse. Fruit a berry, ovoid, greenish, fleshy. tawny-tomentose with out; pedicel stout; seeds 2.

USES:

Leaf paste is applied on head for dandruff and skin diseases. Oil prepared from leaf juice is applied on head for proper growth of hair and to prevent hair fall. Leaves ground in milk or rice washed water is given for four days from menses as a blood purifier and for conception. Leaf or stem decoction is given for jaundice and blood in stool.

Mahonia leschenaultia (Wallich ex Wight & Arn.) Takeda

Takeda Mullu Kadumbuor, Mullu Manjanathi

It is an erect or climbing shrubs, unarmed. Leaves alternate, simple or imparipinnate, membranous, herbaceous, coriaceous, petiolate or sessile, exstipulate, rachis articulated at each leaflet node, leaflets few to 41, sessile to petiolulate. Inflorescences solitary, fascicled racemose or panicle, terminal cymes. Flowers bisexual, yellow or greenish yellow, actinomorphic, pedicellate, sepals 9, usually in 3 series, bracts longer or shorter the pedicels, petals usually in one series, base with or without 2 oblong glands. Stamens 6, slightly adnate, anther opening by valves, ovary superior, ellipsoid, ovules few, basal, erect, style short or sessile, persistent in fruit, stigma, Fruits are berries, purple or black, seeds about 1-7, with conspicuous endosperm.

USES:

The stem bark, prepared as a paste, is administered orally to newly delivered women in the Nilgiris to address fever, cold, and other complications during the postnatal period.

Mangifera indica L. Anacardiaceae

Mango, Aam, Mamaram

A glabrous tree up to 15 m tall. Leaves 11-24 × 4-8 cm, oblong, lanceolate, acuminate, coriaceous, shiny and dark green on upper surface. Flowering panicles erect, conspicuous, longer than the leaves, pubescent. Calyx lobes ovate, pubescent on the outside. Petals imbricate, oblong, inner surface prominently 3-nerved. Drupe ovoid in outline, compressed, 3.5-20 cm long. Mesocarp fleshy. Endocarp (stone) hard and fibrous.

USES:

Seed kernel ground in butter milk is applied on head for dandruff. Bark juice is mixed with white egg is used for dysentery. Bark decoction is recommended for acidity, allergy and intestinal worms.

Manilkara zapota (L.) P. Royen Sapotaceae

Sapota, Chikoo, Sappotta

Small evergreen tree up to 5 m tall. Leaves simple, crowded at the end of branches, ovate-elliptic to elliptic-lanceolate, 2.5-3.5 × 7-12.0 cm, glabrous, subcoriaceous, midrib prominent. Flowers white, 1-1.5 cm across, long pedicelled. Sepals unequal, 3 larger than the others Corolla greenish white, urceolate, lobes 6.. Stamens 6, opposite the petals; staminodes as many as the petals, petaloid. Fruit usually a globose fleshy berry, 5-10 cm in diameter, epicarp thin, rough rusty brown. Seeds 5-12, shining black, obovate, ca. 2.0 cm long.

USES:

Unripe fruit is ground and given for blood dysentery. It helps in preventing colon, lung and oral cavity cancers. The leaves has anti-hyperglycemic and hypo-cholesterolemic activity. The bark is used to treat gastrointestinal disorders and inflammatory conditions.

Martynia annua L. Martyniaceae

Tiger-claw, Thael kodukkukai

Erect subshrubs to 1 m tall, glandular-hairy all over. Leaves 8-20 × 6-19 cm, broadly ovate to suborbicular, base cordate, margin sinuate-dentate, apex acute, chartaceous, palmately 5-nerved from base; petiole 8-14 cm long. Racemes 7-14- flowered, at the bifurcations of the branches; peduncle 4-8 cm long; pedicel 1-2 cm long; bracts 1-2 × 0.7-1 cm, broadly ovate, pale pink. Calyx lobes 5, greenish-white, 9-12 mm long; ovate-oblong. Corolla reddish-purple with yellow throat; tube 3-4 cm long; lobes rounded. Stamens 2; filaments white, 1 cm long; anther cells disjunct, superposed, longitudinally coherent with the opposite pair; staminodes 3. Fruits drupaceous, ovoid, flattened somewhat dorso-ventrally, ribbed, viscid-pubescent; endocarp 2-3 cm long, hard, black, with a hooked spine at tip.

USES:

It is primarily used in the treatment of epilepsy, joint pain, sore throat and snakebites.

Mentha spicata L. (*Mentha viridis* (L.) L.) Lamiaceae

Spearmint, Podina

Herbs stoloniferous. Stems erect, 40 cm, glabrous or subglabrous, striate. Leaves sessile or subsessile, ovate-oblong to oblong-lanceolate, 3-7 × 1-2 cm, base broadly cuneate to rounded, margin irregularly serrate, apex acute. Verticillasters in cylindric terminal spikes 4-10 cm, spikes interrupted at base; bracteoles linear, 5-8 mm. Pedicel ca. 2 mm. Calyx campanulate, ca. 2 mm, glandular, obscurely 5-veined; teeth triangular- lanceolate, ca. 1 mm. Corolla purplish, ca. 4 mm, glabrous, tube ca. 2 mm; lobes subequal, apex emarginate. Ovary brown, glabrous.

USES:

The plant is used to treat gastrointestinal, respiratory, bad breath, carminative, antispasmodic, diuretic and sedative agents.

Magnolia champaca (L.) Baill. ex Pierre (Michelia champaca L.) Magnoliaceae

Champak, Shenbaga

Medium trees, 10m tall. Leaves 13-25 cm long, 4.5-8.5 cm broad, lanceolate to ovate-lanceolate, sparsely rusty hairy on both sides, apex acuminate, base acute; stipules densely rusty-pubescent. Flowers solitary, axillary, 5-7 cm in diameter, fragrant; bracts coriaceous, with rusty or golden tinged hairs; pedicel 3-5 mm long. Perianth of 5-8 segments each, 2.3-5 cm long, 0.3-1.2 cm broad, subequal, pale yellow to yellow. Stamens 5-7 mm long. Carpels many, free; ovary 4 mm long, unilocular, 2–several ovuled. Fruiting axis 5-10 cm long. Follicles 1-2 cm long, nearly as broad as long, brown, sparsely hairy and covered with warty excrescences, dehiscing from the top downwards. Seeds 1-2 or several, 5-12 mm long, brown, polished, usually rounded on the back.

USES:

Bark decoction is given to abort the foetus up to 45 days old. Seed oil is applied for rheumatism. Bark decoction is taken internally to regulate menstruation. This plant is used traditionally to treat colds, coughs digestive issues and pains.

Melia azedarach L.

Meliaceae Persian lilac, Bakain, Malaivembu

A deciduous tree, to 20 m tall... Leaves odd-pinnate, 2-pinnate or 3-pinnate, 20-40 cm; leaflets opposite; leaflet blades ovate, elliptic, or lanceolate, 3-7 × 2-3 cm but terminal one usually slightly larger, secondary veins 12-16 on each side of midvein, cuneate to broadly cuneate, margin crenate or sometimes entire, apex shortly acuminate. Flowers fragrant. Calyx 5-parted; sepals ovate to oblong-ovate, outside puberulent, apex acute. Petals lilac-coloured, obovate-spatulate, 0.9-1.3 cm, both surfaces puberulent but usually outside more densely so. Stamina! tube purple, 7-8 mm, with longitudinal stripes, glabrous or subglabrous, apical margin with ten narrow lobes; lobes conic, further 2- or 3- lobed; anthers 10, inserted on inner side ofl obes. Ovary spherical, glabrous, 5-8-locular, with two ovules per locule; style acerose; stigma capitate, not included within filament tube, apex 5-dentate. Drupe globose to ellipsoid, 1-3 × 0.8-1.5 cm; endocarp ligneous. Seed ellipsoid, smooth, brown, ca. 3.5 × 1.6 mm.

USES:

It is used to treat a wide range of disease such as diarrhea, malaria, and various skin diseases.

Merremia tridentata (L.) Hall. f. Convolvulaceae

Kidney Leaf Morning Glory, Prasarini, Mudiyor kondal

A slender perennial prostrate herb with angular stems. Leaves, various in shape and size, but deeply emarginated and 3-toothed at the apex. Cordate at base, the basal lobes clasping the stem, petioles very short or 0. Inflorescence, Axillary solitary. Flowers, Calyx infundibuliform, outer sepals shorter than inner. Corolla pale yellow. Fruit, pale brown, glabrous, seeds trigonus, glabrous.

USES:

The whole plant is used in rheumatism, kidney stones, and herpes. Boiled and mashed leaves applied on the abdomen in cases of retention of urine. Roots useful in piles, inflammation of spleen, pain in the chest and liver.

Mimosa pudica L. Mimosaceae

Sensitive plant, Lajwanti, Thootasinugi

It is a low-growing, much-branched, prickly, perennial shrub grows 50 cm tall. The leaves bipinnate. Pinnae are in two pairs. The petiole is about 2.5 cm long and the pinnae are 2.5-5 cm long. There are 12-50 leaflets, each 6-12 mm long, 1.5 mm wide, oblong-linear and pointed, glabrous above but with a hairy margin and lower surface. The stipules are linear-lanceolate and 7-8 mm long. Leaflets and leaves fold up rapidly when touched and also close at night. Flowers are bright purplish-pink with more prominent stamens and occur in globular or ovoid heads about 9 mm in diameter. The calyx is minute and the corolla is four-lobed and about 2 mm long. Peduncles, 12-25 mm long, arise from leaf axils, are densely hairy and carry prickles. Fruits are borne in clusters in the leaf axils. Each fruit is an oblong, flattened, recurved pod about 8-20 mm long and 2-6 mm wide containing 1-5 seeds.

USES:

Whole plant decoction is used as gargle for toothache. Leaf juice is applied for cuts and wounds. Thick paste of whole plant with turmeric is applied for three hours to expel glass pieces, thorns, spines from body, also to heal the wounds. Plant paste is applied for rashes.

Mimusops elengi L. Sapotaceae

Spanish tree, Bakul, Magizham

A small evergreen tree 7.5 m high, Stem cylindrical, bark dark grey, smooth, not deeply furrowed. Leaves alternate, broadly ovate, 3.5-4-5 × 7-10.5 cm, base cuneate, apex acuminate, entire or slightly wavy, glabrous, petiolate, petiole 2-3.5 cm long. Flowers axillary solitary or fascicled, creamish white, fragrant, c. 15 mm in diameter, pedicel c. 2 cm long, hairy. Calyx 6-lobed, lobes free, lanceolate, tomentose. Corolla- tube very short, lobes 24 in 2 series. Stamens 8, anthers subsessile, staminodes hairy. Ovary pubescent. Fruit ovoid, 2.5-4 cm long, orange, pubescent when young, with persistent calyx. Seed are egg-shaped, grayish-brown and shining.

USES:

The bark is astringent, bitter and tonic. It is used in the treatment of diarrhoea and dysentery. A decoction of the bark, sometimes mixed with the flowers, is used as a gargle to treat gum inflammation, toothache. It is also used to treat gonorrhoea, scabies and eczema.

Momordica charantia L. Cucurbitaceae

Bittergourd, Karela, Pavakaai

Climbers; tendrils simple. Leaves up to 8 × 8 cm, orbicular-cordate, 5-7-lobed; lobes acute or acuminate, margin dentate-lobulate; petiole to 4 cm. Flowers axillary, solitary. Male flowers: bracts 5-12 × 5-12 mm; foliaceous, reniform; peduncle to 6 cm long; calyx tube campanulate, glabrescent; lobes ovate; Corolla yellow; to 3 cm across, petals 5, obovate, villous. Stamens 3; anthers to 2.5 mm; pistillodes 3. Female flowers: Peduncles 6-10 cm long often bracteate at base; ovary fusiform. Fruit 10-15 cm long, oblong, echinate. Seeds 1-1.5 × 0.7-0.9 cm, ovoid, compressed, base and apex subtridentate, sculptured.

USES:

It is used to treat in lower blood sugar levels, improve insulin levels, and protect pancreatic cells.aiding in the treatment of digestive issues, jaundice, skin infections and hypertension.

Morinda pubescens J. E. Smith Rubiaceae

Morinda tree, Aal, Manjanathi

Deciduous trees, to 12 m high Leaves simple, opposite, decussate; stipules interpetiolar, connate, sheathing, bifid at apex; petiole 5-20 mm long, stout, pubescent, grooved above, glabrous; lamina 7.5-25 × 2-8 cm, elliptic-lanceolate, oblanceolate or elliptic-oblong, base acute, cuneate or attenuate, apex acute to acuminate, margin entire, pubescent or tomentose, chartaceous; lateral nerves 7-12 pairs, pinnate, prominent. Flowers bisexual, white, in terminal, globose heads. Calyx limb truncate. Corolla 2 cm long; tube 1 mm long, villous within; lobes 4, oblong, recurved. Stamens 4, attached to the throat of corolla; anthers exserted. Ovary 1.5 mm, 2-celled or incompletely 4-celled, inferior; style 4 mm; stigma 2 lobed. Fruit a syncarp, 15-18 mm across, globose, pyrenes 4, oblong; seeds not winged.

USES:

Leaf paste is applied for eczema, itches, ulcers and swellings. Bark decoction is recommended internally for fever and digestive disorders.

Morinda citrifolia L. Rubiaceae

Noni, Bartundi, Nuna Ven-Nuna

Small trees or shrubs 3-6 m tall; stems 4-angled, glabrous. Leaves glossy, membranous, elliptic to elliptic-ovate, 20-45 cm long, 7-25 cm wide, glabrous, petioles stout, 1.5-2 cm long, stipules connate or distinct, 10-12 mm long, apex entire or 2-3lobed. Flowers perfect, ca. 75-90 in ovoid to globose heads, peduncles 10-30 mm long, Calyx a truncate rim. Corolla white, 5-lobed, the tube greenish white, 7-9 mm long, the lobes oblong-deltate, ca. 7 mm long. Stamens 5, scarcely exserted; style ca. 15 mm long. Syncarp yellowish white, fleshy, 5-10 cm long, ca. 3-4 cm in diameter, soft and foetid when ripe. Seeds with a distinct air chamber.

USES:

Fruits are used in curing diabetes, high blood pressure, aches, pains, burns, arthritis, inflammation and tumors.

Moringa oleifera Lam. (*Moringa pterygosperma* Gaertn.)

Moringaceae Drumstick, Senjana, Murungai

It is a fast-growing, deciduous tree. It can reach a height of 10-12 m and the trunk can reach a diameter of 45 cm. The bark has a whitish-grey colour and is surrounded by thick cork. Flowers fragrant, bisexual, surrounded by five unequal, thinly veined, yellowish-white petals. Flowers a1.0-1.5 cm long and 2.0 cm broad, grow on slender, hairy stalks in spreading or drooping later flower clusters which have a length of 10-25 cm. The fruit is a hanging, three-sided brown capsule of 20-45 cm size which holds dark brown, globular seeds with a diameter around 1 cm. The seeds have three whitish papery wings and are dispersed by wind and water.

USES:

The leaf extract used to help treat some stomach disorders, such as constipation, gastritis, and ulcerative colitis. It has many important vitamins and minerals. The leaves have 7 times more vitamin C than oranges and 15 times more potassium than bananas. It also has calcium, protein, iron, and amino acids, which help your body heal and build muscles.

Morus alba L. Moraceae

White mulberry, Shahtoot, Kambli Chedi, Mussukottai

A deciduous tall tree up to 5 m tall tree with a dense, compact leafy crown. Leaves with a crisped hairy, filiform, 1-3 (-3.5) cm long petiole. Lamina narrow to broad, ovate, 3- costate from truncate to shallowly cordate base, upper surface glabrous, midrib and principal wins pubescent, margin regularly serrate or crenate-serrate or irregularly 2- serrate, apex obtuse, acute or shortly acuminate. Male catkins 10-20 (-30) mm long including slender, hairy peduncle, with lax flowers. Male flowers: Sepals free, broadly ovate, ca. 2.5 mm long, cucullate, obtuse, glabrous to hairy; staminal filaments equal to sepals, with ovate, exserted anthers. Female catkins ovoid, (2) 5-10 (-12) mm long, with out equally long or slightly longer peduncle. Female flowers: Sepals suborbicular, c. as long as or slightly larger than of male flowers, glabrous or ciliate on margins; ovary with glabrous free styles. Sorosis ovoid, 5-8 mm across, white to pinkish-purple or black, sweet and edible.

USES:

Used in the treatment of dizziness, insomnia, premature aging, and Myotonic Dystrophy type 2.

Mucuna pruriens Baker Fabaceae

Velvet Bean, Kaunch, Poonaikaali

Slightly woody annual vine, twining, attaining 5-7 m in length. Stems pubescent, cylindrical, striate, glabrescent. Leaves alternate, trifoliolate; terminal leaflet rhombic- ovate, lateral leaflets markedly asymmetrical, one side oblong, the other ovate, petioles 4.5-16.5 cm long, strigose, with the base swollen; stipules minute, caducous; stipels linear, 4-5 mm long. Inflorescences of axillary pseudoracemes, 10-30 cm long, with 1-3 flowers per node. Calyx grayish green, campanulate-asymmetrical, 1-1.5 cm long, sericeous- pubescent, mixed with some stiff hairs, the sepals unequal, triangular, deep; standard ovate, 1.5-2 cm long, violet, the wings purple, oblong, ca. 4 cm long, the keel pale violet; staminal column almost white, exserted, the anthers violet. Pod more or less cylindrical, with the curved extremities forming an ‘s’, 4-10 × 1-1.4 cm, densely covered with stiff, extremely stinging hairs.

USES:

It is used in the treatment of Parkinson’s disease, male infertility and nervous system disorders. It also serves as an aphrodisiac, neuroprotective, and treating conditions like low libido, stress and diabetes.

Musa paradisiaca L. Musaceae

Banana, Kela, Vazhai

A large tree-like herb with thick rhizome, pseudostem fleshy, succulent formed by the imbricate leaf sheaths. Leaves large, oblong, petioles long channeled, bright glossy green. Flowers on recurved large, spadix drooping, the lower flowers all female, the upper all male, clustered and enclosed in the axils of large, reddish-purple caduceus, boat-shaped spathes or bracts. Tepals 6, biseriate, 3 + 3 synpetalous in two whorls of 3 each and homochlamydeous. Stamens 5 perfect and 1 small rudimentary or 0, filaments long stout, anthers linear, 2-celled. Ovary inferior, 3-celled, style long, stigma lobulate. Berry, fleshy, narrow at both ends, seeds rarely present in cultivated variety.

USES:

The flower of this plant is used to treat ulcers, dysentery, and bronchitis. Cooked flowers are good for diabetics. The astringent ashes of the unripe banana peel and leaves are used in the treatment of dysentery and diarrhea and also for the treatment of malignant ulcers.

Myristica fragrans Houtt. Myristicaceae

Nutmug, Jaiphal, Jathikkai

A dioecious evergreen tree, 7.5 m tall. Leaves alternate, simple, exstipulate, chartaceous; petiole about 1 cm long; blade elliptical to lanceolate, 5-15 cm × 3-7 cm, base acute, margin entire, apex acuminate, Flowers fragrant, pale yellow, with a 3 lobed perianth; male flowers with a slender pedicel less than 1 mm thick. female flowers with a superior, sessile, 1 celled ovary with a single basal ovule. Fruit peach shaped, berry or drupe-like, 5-8 cm long, fleshy, yellowish, splitting open into 2 halves when ripe, containing 1 seed.

USES:

The seed oil a bitter, astringent and serves sedative to treat insomnia and to improve cognitive function. It controls vomiting and relaxes spasms. The seed is also said to have stimulant, carminative and aphrodisiac properties.

Naravelia zeylanica (L.) DC Ranunculaceae

Ceylon Clematis, Vatanasini, Vatamkolli

Climbing shrubs; young stem, leaves and buds densely villous; roots tuberous. Leaves 2-3-foliolate; leaflets 4-12 × 3-8 cm, elliptic-ovate, base rounded, margins distantly dentate, apex acute or acuminate, pubescent beneath, glabrous above, basally 5-ribbed; terminal leaflet transformed into a 3-fid, hooked tendril, up to 9 cm long; petioles 5-10 cm long. Flowers c. 1.5 cm across, bisexual, in terminal and axillary, divaricately branched, up to 15 cm long panicles. Sepals 4-5, greenish yellow, 0.8-1 cm long, elliptic, pubescent without. Petals 6-12, greenish yellow, 0.7-0.9 × 1-1.5 mm, linear to spathulate. Stamens many; filaments ligulate; staminodes 10-14, petaloid. Carpels many; ovule 1per carpel; style 1.5-2 mm long; stigma clavate. Achenes many; 0.8-1 cm long, linear, stalked, with spirally twisted, 3-4 cm long, feathery persistent style.

USES:

The plant is used in the treatment of helminthiasis, dermatopathy, leprosy, rheumatalgia, odontalgia, wounds and ulcers.

Naringi crenulata (Roxb.) Nicolson Rutaceae

Leaf Limonia, Beli, Mahavilvam

Trees, to 10 m, armed with sharp solitary or paired axillary spines; bark yellowish-grey, smooth, corky; blaze yellow; branchlets angled, glabrous. Leaves imparipinnate, alternate, estipulate; rachis 2.5-10 cm long, winged, wings obovateoblong, glabrous, punctate; leaflets 3-7, opposite, sessile, estipellate; lamina 1-6 × 0. 7-2.4 cm, elliptic or elliptic-obovate, base acute or obliquely acute, apex obtuse to emarginate, margin crenate, glabrous, profusely pellucid-gland dotted; glands dimorphic; coriaceous; lateral nerves 4-10 pairs, pinnate, slender, faint, intercostae reticulate, faint. Flowers bisexual, white, in few flowered axillary racemes; pedicel 8-10 mm long; sepals 4, free or united at base, ovate-orbicular, glandular; petals 4, free, elliptic or oblong, white, glandular, glabrous; stamens 8, subequal, free, inserted round the disc; filaments subulate; anthers yellow, apiculate, disc thin, glandular; ovary superior, globose, 4-celled, glabrous, glandular, ovule 1 in each cell.; style stout; stigma capitate; Fruit a berry, 6-8 mm across, subglobose, bluish-black, glandular; seeds 1-4, dull yellow, smooth.

USES:

The plant root is used to curing vomiting, dysentery and colic disorders. Leaves are used to lower blood sugar levels and protect the liver from toxic injury.

Nelumbo nucifera Gaertn. Nymphaeaceae

Gaertn Indian Lotus, Kamal, Thamarai

An aquatic perennial herb, rhizomatous, grows up to a height of ca. 150 cm and horiziontal spread of up to 3 m. Leaves peltate, orbicular, 60 – 90 cm, glabrous, glaucous, dark green above, pale beneath, covered with a network of microscopic hairs; veins radially extended; petioles very long, rise above the water, smooth or minutely prickled. Flower solitary, to 25 cm in diam., bisexual, entomophilous. Sepals elliptic or ovate, 1.5 – 5 × 1 – 3.5 cm, concave, green or pinkish green. Petals ca. 25 (in single form) and ca. 110 (in double form), elliptic, obovate to spathulate, obtuse or subacute, 4 – 15 × 2 – 8 cm, concave, rose-pink or white. Stamens numerous. Receptacles 2 – 4 cm across, spongy. Carpels numerous, loosely embedded in cavities on flattened top of receptacle, 1-ovuled. Fruit a nut-like achene, oblong to ovoid.

USES:

The Roots help in removal of toxic wastes from body, and also helpful in reducing body heat. The cooked root is good for the stomach and the reproductive organs. Rhizome paste is applied in ringworm and other cutaneous affections. Stem helps in the healthy growth of the foetus. Leaf paste is applied to the body in fever and inflammatory skin conditions. Leaves and flowers are useful in many bleeding disorders. Flowers are prescribed to promote conception.

Neolamarckia cadamba (Roxb.)Bosser Rubiaceae

Burflower tree, Kadam, Vellaikadambu

Deciduous trees, to 20 m high, bark 6-8 mm, pale brown, vertically shallowly grooved, exfoliating in small rectangular flakes; blaze yellowish-brown; bole straight; branches horizontal. Leaves opposite, decussate; stipules interpetiolar, lanceolate, cauducous; petiole 20-45 mm, stout, glabrous; lamina 10-25 × 6-12 cm, ovate or elliptic- oblong, base truncate or obtuse, apex acuminate, margin entire, glabrous above, pubescent beneath; lateral nerves 10-14 pairs, pinnate, prominent; intercostae scalariform, prominent. Flowers bisexual, yellowish, in globose heads, 2-4.5 cm across; calyx tube 2-3 mm in diameter, tube globose, lobes 5, 5-6 × 3-4 mm, membranous; corolla tube 6-8 mm long, lobes 5, oblong, acute, 3-4 cm long, glabrous; stamens 5, 3 mm long, anthers sagittate, sessile; ovary 3-4 mm long, 2-celled at base, 4-celled above, inferior, ovules many; style exserted to 5-6 mm, entire; stigma clavate. Fruit a capsule on a fleshy globose receptacle, 3.5-5 cm across, orange yellow, capsule membranous; seeds many, angular, minute.

USES:

It is used in the treatment of various ailments such as fever, uterine complaints, blood diseases, skin diseases, tumour, anaemia, eye inflammation and diarrhoea.

Nothapodytes nimmoniana (Grah.)Mabberley Icacinaceae

Stinking Tree, Ghenera, Arali

Small trees, to 5 m high, bark yellowish, wrinkled; branchlets angled, appressed- hairy, corky, with prominent leaf scars. Leaves simple, alternate, estipulate; petiole 30-60 mm, stout, pubescent, grooved above; lamina 7-18 × 4-8 cm, broadly ovate, ovate-oblong or elliptic-oblong, base oblique, round or acute, apex acuminate, margin entire, glabrous above, pubescent beneath, coriaceous; lateral nerves 6-10 pairs, pinnate, prominent, intercostae scalariform; domatia present. Flowers bisexual, creamy yellow, 5 mm across, in terminal pubescent corymbose cymes or panicles; calyx cupular, 1 mm long, 5-toothed; petals 5, 3 mm long, villous within, apically inflexed; stamens 5, free; disc cupular; ovary ovoid, hairy, 1-celled, ovules 2; style angled; stigma stout. Fruit a drupe, 2 × 1 cm, ellipsoid, blackish-purple; seed glabrous.

USES:

It is used to treat sarcomas, including lung, breast, and uterine cervical cancers. Also to treat mental illness, skin diseases, tuberculosis, anemia, jaundice, and hypertension.

Nyctanthes arbor-tristis L. Oleaceae

Cooral Jasmine, Parijat, Pavizhamalli

It is a deciduous shrub up to 4 m tall, with quadrangular branches and grey or greenish-white rough bark. Leaves are simple petiolate and exstipulate, opposite, 3-10 by 4-6.3 cm ovate, acute or acuminate, bears distant teeth on entire, short bulbous hairs rounded or slight cuneate, petiole 6cm long hairy. The flowers are originated at the tips of branches in the axils of leaves and are small in clusters of 5-7 fragrant delightfully, peduncles of 4 angled, slender. Calyx are 6-8 mm long, narrow campanulate, hairy glabrous inside, truncate toothed or lobed, ciliated. Corolla glabrous more than 13mm long, 6-8 mm long tube, orange colour, equaling limb, white lobes corolla with an orange red center, unequally obcordate. Fruits are flat brown heart shaped to round capsule with two sections each containing a single seed.

USES:

This plant is used in various ailments like fever, enlargement of the spleen, malaria, blood dysentery, cough and gastritis. Juices of leaves is used as digestives, antidote to reptile venom Seeds are used to cure scurfy infection of scalp, piles and skin diseases. Powdered stem bark is given in rheumatic joint pain. Oil is used for pain in the eyes and with Arjuna bark, it is rubbed on the body for internal injury.

Nymphea pubescens Willd

Wild Water Lily, Koka, Ambal

Perennial floating herbs with rhizomatous stem. Leaves up to 28 × 26 cm, orbicular, sagittate to cordate, sharply sinuate-toothed densely hairy on the lowerside, sparcely hairy to granulated on the upper surface, petiole up to 1 m and pedicel up to 70 cm long. Flowers about 8 cm across. Sepals 3.5-7 × 1.5-2.5 cm, oblong-oblong lanceolate, entire, obtuse to acute, 5-7 ribbed, greenish. Petals 4-7 × 1.5-3 cm, whitish, bluish with pinkish tinge, oblong, ovate-oblong. Stamens in 3 rows, 1.5-35 cm long. Carpels (stigmatic rays) 15-30, club shaped, converging towards the centre of the flower. Fruit 3.5-5 cm, globose.

USES:

Rootstock is used to treat urinary ailments, cystitis, nephritis, and haemorrhoids. Flower acts as an aphrodisiac, blood purifier and is used to treat palpitation of the heart.

Ocimum americanum L.

Lemon Basil, Bapchi, Naiithulasi

An aromatic annual herb up to 45 cm tall. Stems 4-angled, woody at base… Petiole 5-12 mm, puberulent and pilose; leaf blade ovate, 1-1.7 cm × 5-10 mm, puberulent, glandular, base cuneate, margin serrulate to subentire, ciliate, apex acute, Racemes numerous, erect, 10-14 cm, pedunculate, verticillasters separate, 6-flowered; Pedicel ca. 1.5 mm, conspicuously recurved in fruit. Calyx campanulate, ca. 2 × 1 mm, puberulent near base, throat densely puberulent, tube ca. 1.5 mm; middle tooth of upper lip circular, lateral teeth shorter, triangular; lower lip teeth ovate-lanceolate, apex awned, longer than upper lip, ciliate; fruiting calyx ca. 4 mm, conspicuously 10-veined. Corolla straight, as long as calyx, limb sparsely puberulent, glabrous inside; tube conspicuously exserted, obconical; lower lip oblong, concave, margin entire. Stamens free, Nutlets black, ovoid, ca. 1.5 × 0.6 mm, finely dotted.

USES:

Decoctions are used for cough. Pounded leaves are placed on the forehead to relieve catarrh or on the chest for respiratory problems; The whole plant is used in baths to treat rheumatism, renal colic and calcification. The plant has been listed as a potential medicine against cancer.

Ocimum basilicum L. Lamiaceae

Sweet Basil, Babul, Thiruneetrupachilai

A small, much branched under shrub with square stem up to o.75 m tall. Leaves opposite, ovate 2 cm long, acute at tip, gland –dotted, borne on thyrses. Calyx bilabiate, hairy. Corolla bilabiate, pink or purple with long white hairs. Stamens 4, didynamous. Ovary bicarpellary, 4-celled, deeply 4-lobed, style gynobasic, stigma bifid. Nutlets 4 with persistent calyx.

USES:

The leaves are used in decoctions to treat cough, cold, bronchitis and fever. Leaves are also applied to treat skin diseases, itching, warts and insect bites. Seeds act as a carminative and helps to manage diarrhea, dysentery and vomiting.

Ocimum gratissimum L. Lamiaceae

Clove Basil, Ram Tulsi, Perunthulasi

Perennial shrubs 1-2 m tall; Leaves opposite; petiole 2-4.5 cm long, slender, pubescent; blade elliptical to ovate, 5-16 × 3-9 cm, membranaceous, sometimes glandular punctate, base cuneate, entire, margin elsewhere coarsely crenate- serrate, apex acute, puberulent or pubescent. Inflorescence a terminal, simple or branched raceme 5-30 cm long; flowers in 6-10-flowered verticillasters, small. Calyx 2-lipped, 2-3 mm long, in fruit 5-6 mm, pubescent, upper lip rounded and recurved. Corolla campanulate, 3.5-5 mm long, 2-lipped, greenish-white, pubescent outside, upper lip truncate, 4-fid, lower lip longer. Stamens 4 in 2 pairs, inserted on the corolla tube. Ovary superior, carpels, 2 each 2-celled, style 2-fid. Fruit consisting of 4, dry, I-seeded nutlets enclosed in the persistent calyx; nutlet subglobose, 1.5 mm long, rugose, brown.

USES:

The whole plant is used to treat stomach aches, diarrhea, vomiting and aids indigestion. It is also used in the treatment of diabetes, cancer and inflammation. The essential oil is used to treat fever, headache and pneumonia

Ocimum tenuiflorum L. Lamiaceae

Holy Basil, Syama Tulasi, Tulasi

An erect, much branched, perennial herb up to 75 cm tall, a pungent aromatic odour. Leaves with blades elliptic to elliptic-oblong, 3-6 cm long, 1-2.5 cm wide, cuneate to attenuate at base, obtuse to acute at apex, entire to remotely serrate at margins. Petioles 1-2.5 cm long, softly pubescent. Flowers terminal, slender racemes or panicles 4-12 cm long, 1-1.5 cm wide, the bracteoles 2-3 mm long, ovate, acuminate, ciliate; flowers in verticils. Calyx c 2.5 mm long at anthesis, in fruit up to 5 mm long, glabrous within, the upper lip suborbicular, reflexed, short-apiculate, the lower lip longer than the upper lip, the teeth 4, lanceolate. Corolla pale pink, pale lavender or white, to 4 mm long. Filaments of stamens exserted, slender, the upper pair of each with a small, bearded basal appendage. Fruit with nutlets purple-green to brown, broadly ellipsoid, 0.8-1.2 mm long, smooth to minutely pitted, swelling in water.

USES:

The juice of leaves possesses diaphoretic, antiperiodic, stimulating and expectorant properties. It is used in catarrh and bronchitis, applied to the skin, in ringworm and other cutaneous diseases and dropped into the ear to relieve ear ache. An infusion of the leaves is used as a stomachic in gastric disorders of children. The seeds are mucilaginous and demulcent, and are given in disorders of genitourinary systems.

Operculina turpethum (L.) Silva Manso

Indian Jalap, Nisoth, Sivathai

Large, perennial twiner up to 10 m long. Leaves alternate, simple and entire; blade very variable in shape, orbicular, broadly ovate to lanceolate, 5.5-15 cm × 1-14 cm, base cordate to hastate, apex acuminate, acute or obtuse, mucronulate, margin rarely shallowly lobed, with 8-11 pairs of veins; major veins prominent below. Inflorescence a few- flowered axillary cyme. Flowers 5-merous. Pedicel 1.5-3.5 cm long, pubescent, in fruit club-shaped, up to 40 mm long; bracts oblong or elliptical-oblong, 1.5-2 cm long. Sepals ovate to broadly ovate, acute, outer ones 1.5-2.5 cm long, pubescent outside, inner ones c. 2 cm long, glabrous, calyx in fruit broadly cup-shaped, up to 6 cm in diameter. Corolla broadly funnel-shaped, 3-4.5 cm long, white, pinkish or yellowish, glabrous. Stamens 5, inserted, filaments adnate to the corolla. Ovary superior, 2-celled, glabrous, style c. 1 cm long, stigmas 2, capitate. Fruit a depressed-globose capsule, c. 1.5 cm in diameter, with 4 seeds.

USES:

An infusion of the rhizome is taken internally to treat fever, anorexia, obesity, oedema, hepatosplenomegaly, hepatitis, intoxication and abdominal tumours. Externally a paste of the rhizome is applied to treat scorpion stings, snakebites, ulcers, worm infestation, wounds, pruritus, vitiligo and other skin disorders. Oil extracted from the rhizome bark is used in skin diseases with scale formation. Rhizome powder mixed with butter and honey is taken to treat vomiting of blood, tuberculosis and herpes.

Ormocarpum cochinsinensis (Lour.) Merr. Fabaceae

Cochinchina plant, Duthiya Choti, Elumbotti

Erect subshrubs. Leaves pinnately 10-13-foliolate; leaflets alternate, 2-3 × 1.2 cm, obovate-oblong, obtuse; petiole slender; stipule ovate. Flowers in slender axillary, 6-10 cm long, racemes; bracts ovate; calyx 6 mm long, lobes ovate, acute, hairy; corolla 8 mm, long, pink; standard clawed, wings 5 × 4 mm, orbicular, auricled; stamens monadelphous, spitting later; ovary linear, many-ovuled, hispid. Pods 2-3 cm long, 1-4 jointed; joints oblong with soft echinate process.

USES:

It is traditionally used for curing bone fracture and used as an anti - inflammatory agent.

Oxalis corniculata L. Oxalidaceae

Indian Sorrel, Changeri, Puliyarai

It is perennial with a slender primary root. Stems are prostrate to suberect, to about 40 cm long, slender, weak, branched, often rooting at nodes, covered with spreading flexible hairs. Leaves with 3 leaflets. Lamina of leaflets equal, 5-18 × 7-23 mm, broadly obcordate, glabrous or sparsely hairy, with densely hairy midrib below. Sepals 3-4.5 mm long, elliptic-lanceolate, elliptic-ovate to elliptic-obovate, with forward-pointing hairs; margins sometimes red; calli 0. Petals 4.5-10 mm long, oblong-obovate, yellow, glabrous. Stamens at 2 levels; filaments glabrous, connate towards base. Styles less than, equal to or a little longer than the stamens, densely hairy. Capsule (6)-10-18-(20) mm long, excluding styles, cylindrical, usually densely clothed in short, simple, more or less backward- pointing hairs, or sometimes hairs sparser and then also with scattered, long, wide, curly, septate hairs, occasionally glabrous.

USES:

Leaves acts as an apptizer, improves digestion and helps to treat diarrhea and abdominal pain. It is also used to treat fever, cold, urinary track infections and high blood pressure. Leaves and fruits are edible and rich in vitamin ‘c’.

Pandanus amaryllifolius Roxb. Pandanaceae

Fragrant Pandan, Rampe, Ramba Plant

It is an evergreen tree with fragrantly-scented leaves. It has two distinct growth forms. A small growth form that grows from sucker shoots with slender stems 1-1.6 metres tall and 2-5 cm in diameter. A large growth form that eventually produces an erect stem, 2-4.5 metres tall and 15 cm diameter. Leaves are linear with an entire leaf margin and acute apex. The female inflorescence has not been observed, while the male inflorescence is very rare and only found on the tree form of this species. The male inflorescence is a spike of flowers with a white spathe.

USES:

This plant is having anti-diabetic, anti-inflammatory, and antioxidant properties. Its leaves are used to manage blood sugar, relieve arthritis, reduce fever, treat skin infections, and alleviate scalp issues like dandruff.

Passiflora foetida L. Passifloraceae

Bush Passion, Jhumka Lata, Kattukodithodai

A branched annual or perennial herbaceous vine 1-5 m tall. Leaves single, alternate, stipules to 1 cm long and divided into hair-like segments, petiole 2-10 cm long without nectary glands, blades 5-15 cm long, 3 or 5 lobed, the base cordate, the edges generally fringed with glandular hairs, the veins prominent, pale green and often finely hairy. Tendrils leaf-opposed, unbranched, coiling and grasping. Flowers solitary in upper leaf axils, peduncle 3-5 cm long, bracts 2-4 cm long and deeply divided into hair-like segments that surround the flower and fruit, sepals 5, greenish petals 5, blunt, white to pale purple or pinkish, 3-5 cm across surrounding a 2-rowed corona of purplish filaments, 5 stamens spreading at the top of a column, styles 3. Fruits oval, 2-3 cm long, smooth, enclosed in hairy bracts, atfirst fleshy and green, maturing dry and yellow.

USES:

Plant decoction is widely used to treat anxiety, insomnia and hysteria. It is used to treat coughs, asthama and throat infections. It is also used as a remedy for diarrhea, indigestion and stomach pain. Fruit is emetic and its extract is recommended for asthma and gas trouble.

Pedalium murex L. Pedaliaceae

Large Caltrops, Baragokhru, Yanai Nerinjil

It is an erect or slightly ascending herb, up to 40 cm high, with a glabrous stem. Leaves opposite, ovate-elliptic, 2-5.6 × 2-3.8 cm, cuneate at base, lobed with toothed margins, lobed-emarginate truncate at apex, semi-succulent, glabrous, arranged in 4 pairs, petiole 3.5 cm. Flowers about 3 cm long, yellow, solitary, axillary, with short pedicel. Calyx 5-partite, lobes lanceolate, 0.3 mm long, greenish-purple. Corolla funnelform, tube cylindric, up to 2 cm long, limb slightly 2-lobed; lobes 5, obovate-rounded. Stamens 5, staminode 1; filament slender, unequal, 0.9 cm long, glandular-ciliate; anther 2-celled, 0.1 mm long, glabrous. Ovary 2-celled, globose, 1.5 mm across; 2-loculed; ovules 2; style slender, 1 cm long, glabrous; stigma 2-lobed, 0.5 mm long. Drupe obovoid, 4-angled, hard, 0.1 mm across, 4-spined; seeds 2, oblong, 3.5 mm long.

USES:

Plant extract or seed decoction is used to treat urinary tract disorders, kidney stones and reproductive issues. It is also used as an aphrodisiac, diuretic, for managing blood pressure and to treat digestive disorders, skin diseases & inflammation.

Pergularia daemia (Forssk.) Chiov. Asclepiadaceae

Treelis Vine, Utaran, Velipparuthi

Twining herb with a stem that exudes milky latex, glabrous, pubescent, or tomentose. Leaves opposite, cordate, 4.5-7 × 5-8.5 cm, cordate at base, entire at margin, acute at apex, tomentose, glaucous beneath. Flowers arranged in umbelliform axillary, pedunculate cymes about 17 cm long, individual flowers about 1 cm across, greenish-white. Calyx urceolate, campanulate, 5 elliptic-lanceolate lobes, 3 mm long, rounded at base, ciliate at margin. Corolla-tube short, campanulate, rotate, 5 ovate lobes, 8 mm long, base cuneate, margin densely ciliate-villous, apex acute. Pollinia pendulous; corona two parts: outer flabellate, membranous, imbricate; inner corona 5 mm, alternating with outer, apex 2-fid, acuminate. Ovary globose-obovate, 1.1 mm, 2-loculed; ovules many; style slender 3 mm long, apex slightly convex; stigma small. Follicles oblong, curved, 7 cm long, echinate smooth.

USES:

Leaf juice is used to treat cough, asthma, bronchitis, and catarrhal infections. It is also used for jaundice, liver disorders, and other digestive issues. The plant is effective for rheumatism, arthritis, and muscular pains. Latex is applied to boils, sores, and toothaches.

Phoenix dactlifera L. Arecaceae

Date Palm, Khajur, Perichchai

It is perennial, dioecious tree up to 12 m tall. Leaf pinnately compound, generally more than 7m length; petiole generally armed, bases persistent on trunk; leaflet margins folded upward, proximal leaflets occasionally reduced, spine-like. Inflorescence within crown, more leaves, flowers borne singly. Flower: perianth pale yellow; sepals fused proximally; petals generally free; pistils 3. Drupe generally fleshy sweet. Fruit 2.5-5 cm, oblong-ovate, brown, pulp thick.

USES:

The fruits are used as demulcent, expectorant and laxative. They are used to treat respiratory diseases and fevers. The tree yields a gum that is used in treating diarrhoea. The fruits are having potential health benefits such as chemoprevention of cancer, prevention of diabetes and cardiovascular diseases.

Phyla nodiflora (L.) Greene Verbenaceae

Creeping Lip Plant, Jalputi Jalpapli, Poduthalai

Perennial, prostrate herb with somewhat woody rootstock, rooting at nodes, appressedly pubescent to glabrescent. Leaves oblanceolate, obovate to spathulate, somewhat fleshy, 5-40 mm long, 4-20 mm broad, serrate above, entire below, glabrous to appressedly pubescent, subsessile to sessile, obtuse, rarely subacute. Spikes 1-4.5 cm long, 6-8 mm broad, solitary, axillary, peduncled, appressedly pubescent to glabrous. Flowers very small, white, rarely pinkish, 3 mm long; bracts 2 mm long, mucronate or acuminate, imbricate. Calyx flattened, shorter than bracts, hyaline-membranous, deeply dissected with lanceolate lobes, pubescent. Corolla slightly exceeding the bracts, unequally 4-lobed with spreading lobes. Fruit ovate, 1.6 mm long, subcompressed, enclosed by the persistent calyx, separating at maturity into two, 1-seeded pyrene

USES:

Tender leaf extract is used to treat fever, cough, bronchitis, skin infections. It is also used for digestive disorders and to promote renal health. It helps in removing dandruff.

Phyllanthus amarus Schum. & Thonn. Euphorbiaceae

Stone Breaker, Bhumi Amla, Kizhanelli

An erect herb, up to 60 high; stem cylindric. Leaves alternate, oblong-obovate, 06-08 × 3.5-4.5 mm, rounded at base, entire at margin, obtuse-rounded at base, entire at margin, obtuse at apex, glabrous; 4-pairs beneth; petiole 0.3 mm long. Flowers solitary, axillary, greenish. Male flowers ca. 0.3 mm across, small; pedicel filiform, 04 mm long. Tepals 5, lobe obovate, 0.3 mm long. Stamens 3; filament connate; anther obovoid. Disck of gland 5, minute. Female flowers ca. 01 mm across, minute; pedicel thick, 01 mm long. Tepals 5, lobe obovoid. Ovary globose, 0.4 mm across; 3-loculed; ovules 6; style 3, connate at base, 2-fid; stigma sessile. Capsules depressed-globose, 1.3 mm long; seeds 4, triqutrous, 01 mm across, muricate, minutely hairy, brown.

USES:

Whole plant juice or its decoction with cumin seeds is used for jaundice. Root extract in milk or buttermilk is taken for jaundice. Tender shoot tip decoction is used for blood dysentery. The plant is also widely used to break down urinary and gall bladder stones, acts as diuretic and helps with urinary disorders. It helps in reducing blood glucose levels and improve insulin activity.

Phyllanthus emblica L. Euphorbiaceae

Indian Goosebery, Amla, Nelli

Deciduous trees, to 15 m high, bark grey-brown, rough, irregularly flaking. Leaves simple, alternate, bifarious on short deciduous branchlets, closely overlapping, subsessile, 0.4-1.5 × 2-4 mm, oblong or linear-oblong, apex obtuse and shortly apiculate, base round, glabrous, membranous, nerves obscure; stipules minute, lateral, linear. Flowers unisexual, 2-3 mm across, greenish-yellow, densely clustered in leaf axils. Male flowers: tepals 6, oblanceolate, 1.5 mm, obtuse, stamens 3, anthers oblong, connate by their connectives; apiculate; disc glands 6. Female flowers: tepals 6, oblanceolate, obtuse; ovary superior, 1.5 mm, 3-celled; ovules 2 in each cell; styles 3, broadly fimbriate, recurved, stigmatiferous. Fruit a capsule 1.5-2.5 cm across, subglobose, dehiscing into 6 cocci, disc enlarged to give an appearance of fleshy yellowish-green, indehiscent berry.

USES:

Fruit powder in buttermilk is taken for jaundice, diarrhea, indigestion and bleeding piles. Root bark is ground with fenugreek is recommended for chronic dysentery. Stem bark decoction is given to arrest diarrhoea. Fruit juice mixed with turmeric powder is used for diabetes. It is used in treating colds, cough and bronchitis. Fruits used as a component in Chawanprash and Triphala.

Phyllanthus maderaspatensis L. Euphorbiaceae

Madras Leaf Flower, Kanocha, Melanalli

An erect herb, up to 50 high; stem slender, glabrous. Leaves alternate, obovateoblnaceolate, 1-1.4 × 04-06 cm, cuneate at apex, entire at margin, obtuse-retuse, slightly apiculate at apex, glabrous; obsurely 5-pairs; petiole 1.5 mm long. Stipule narrowly lanceolate, 01 mm long. Flowers solitary, axillary, fascicled. Male flowered ca. 01 mm across, small, greenish. Tepals 6; lobe obovate-oblong, 01 mm across, obtuse at apex. Stamens 3; filament adnate at disk; anther ovoid. Disk with 6-glands. Female flowers ca. 02 mm across, greenish. Tepals 6; lobe suborbicular-obovate, 1.2 mm across, obtuse at apex. Stamens 3; filament adnate at disk; anther ovoid. Disk with 6-glands annular. Ovary 3- lobed, 01 mm across; 3-loculed; ovules 6; style 3, 0.2 mm long; stigma sessile. Capsules circular, rounded, 02 mm across, glabrous; seeds 6, ovoid, 01 mm across, muriculate, black.

USES:

Plant extract in milk is used for jaundice and other liver disorders. Whole plant is used to treat urinary tract infections. Root decoctions are used for constipation, digestive disorders, intestinal pain and diarrhea.

Phyllanthus reticulatus Poir. Euphorbiaceae

Black Honey Shrub, Kale Madhu Ka per, Karunelli

Straggling shrub, with slender, drooping branchlets.; branchlets, hairy-tomentose when young. Leaves alternate, elliptic-oblong to obovate, 1.3-1.7 × 1-1.2, rounded at base, entire at margin, obtuse-rounded at apex, glabrous; 5-pairs; petiole 01 mm long, radish. Stipules deciduous. Flowers ca. 03 mm across, greenish-purple, solitary, axillary, 8-10-flowered; peduncle, up to 05 mm long. Male flowers; calyx-lobe 5, ovate-elliptic, 2.2 mm long. Female flowers; calyx-lobe 5, ovate-suborbicular, 1.5 mm long. Petals absent. Disc in female flower glands 5; male flowers glands 5, small. Stamens 3-5; filament short, thick; anther ovoid, basifixed. Ovary ovoid, 1.5 mm across; 3 or 5-12-loculed; ovules 6; style short, reflexed; stigma didynamous. Berries baccate, globos, 04 mm across, glabrous purple-black when ripe; seeds many, trigonous, angled, 2.2 mm long.

USES:

Leaf boiled in water is used as gargle for softening the gum and gum bleeding. It helps in hair growth. The plant is known to have analgesic, anti inflammatory, cholesterol lowering, immunomodulatory and anti diabetic properties. Also used in gastrointestinal disorders.

Physalis minima L. Solanaceae

Country Gooseberry, Ban Tipariya, Soddakku Thakkali

A straggling shrub with slender, drooping branchlets, hairy-tomentose when young. Leaves alternate, elliptic-oblong to obovate, 1.3-1.7 × 1-1.2 cm, rounded at base, entire at margin, obtuse-rounded at apex, glabrous, 5 pairs of veins, petiole 0.1 mm long, radish-colored. Stipules deciduous. Flowers about 0.3 mm across, greenish-purple, solitary, axillary, 8-10 per peduncle, up to 0.5 mm long. Male flowers: calyx-lobe ovate-elliptic, 2.2 mm long. Female flowers: calyx-lobe ovate-suborbicular, 1.5 mm long, petals absent. Disc in female flowers with 5 glands; male flowers with 5 small glands. Stamens 3-5, filaments thick, anthers ovoid, basifixed. Ovary ovoid, 1.5 mm across, 3 or 5-12 locules, 6 ovules; style short, reflexed; stigma didymous. Berries baccate, globose, 0.4 mm across, glabrous, purple-black when ripe. Seeds many, trigonous-angled, 2.2 mm long.

USES:

Leaf boiled in water is used as gargle for softening of gum and gum bleeding. It helps in hair growth. The plant is known to have analgesic, anti-inflammatory, cholesterol lowering, immunomodulatory and anti-diabetic properties. Also used in gastrointestinal disorders.

Piper longum L. Piperaceae

Long Pepper, Pippali, Valmilagu

Slender undershrubs, leaf blade ovate to reniform, those at apex of stem ovate to ovate-oblong, 6-12 × 3-12 cm, papery, densely glandular, base cordate, basal lobes rounded and equal, slightly incurved, leaf blades toward apex of stem sometimes with basal lobes overlapping, slightly unequal, apex acute to acuminate; veins 7, apical pair partly closely parallel to midvein, reaching leaf apex, others basal; reticulate veins lax. Spikes leaf-opposed, recurved. Male spikes 4-5 cm × ca. 3 mm; peduncle 2-3 cm; bracts suborbicular, sometimes slightly cuneate, ca. 1.5 mm wide, peltate, glabrous, stalk short. Stamens 2; filaments very short; anthers ellipsoid. Female spikes (1-) 1.5-2.5 cm × 2.5-4 mm, 2-3 cm in fruit; peduncle and rachis as in male spikes; bracts 0.9-1 mm in diam. Ovary ovoid, partly connate to rachis; stigmas 3, ovoid, apex acute. Drupe globose, ca. 2 mm in diam., apex umbonate, partly connate to rachis

USES:

Fruits are used with honey for cough. Fruit fried with ghee and salt is given in the morning and rasam prepared from its seeds and salt at night for 11 days after delivery. Fruit ground with water is recommended for speech clearance in children. Root paste is applied for leucoderma. Root ground in butter milk is applied for nervous debility.

Piper nigrum L. Piperaceae

Black Pepper, Kalimirch, Milagu

Climbers woody. Nodes clearly enlarged and rooting, glabrous. Petiole 1-2 cm, glabrous; leaf blade ovate to ovate-oblong, rarely suborbicular, 10-15 × 5-9 cm, thick, leathery, glabrous, base rounded, usually slightly oblique, apex acute; veins 5-7 apical pair arising 1.5-3.5 cm above base, alternate, others basal; reticulate veins prominent. Flowers polygamous, usually monoecious. Spikes leaf-opposed, to as long as leaves; peduncle nearly as long as petioles, glabrous; bracts spatulate-oblong, 3-3.5 × ca. 0.8 mm, adaxially adnate to rachis, only margin and broad, rounded apex free, shallowly cupular. Stamens 2, 1 on each side of ovary; filaments thick, short; anthers reniform. Ovary globose; stigmas 3 or 4, rarely 5. Drupe red when ripe, drying black when unripe, globose, 3-4 mm in diam., sessile.

USES:

Seed paste is applied for rheumatism and decoction for vomiting, to increase appetite and metal poisoning. Oil prepared by boiling leaf juice in gingelly oil is applied for stroke and septic wounds. Fruits fried in ghee are mixed with salt and given for cold and rhinitis. Seeds, clove and lime juice paste is applied for migraine. Fruit ash mixed with bull horn ash and gingelly oil is applied for leprosy and other chronic skin diseases.

Pisonia grandis R. Br.

Lettuce Tree, Kaua Phal, Sandi Keerai

Its is an evergreen tree with an irregular crown, usually growing up to 5 m tall. The leaves are ovate-oblong to oblong, 15-25cm long wide usually unequal obtuse at base acute apex. It has a long petiole with prominent reticulate venation and minutely puberlous in the axils of the nerves. Flowers are small, greysh funnel shaped, decidious with perianth short, 3 mm long.

USES:

The leaves are used in the treatment of dysentery, inflammation, wound healing, algesia and ulcer. Leaves of the cultivar 'Alba' are crushed or heated and applied ton the swellings or open ulcers, corns, calluses, or applied for oedema of the legs.

Plectranthus amboinicus (Lour.) Spreng. Lamiaceae

Mexican Mint, Patharchur, Karpooravalli

Perennial succulent aromatic herbs 50 cm high. Leaves petiolate with blades fleshy, broadly ovate to suborbicular, rhombic, or reniform, 4-10 cm long, 3-9 cm wide, rounded to truncate and then often long-attenuate at base, obtuse to rounded at apex, coarsely crenate to dentate at margins or entire toward base. Flowers terminal, spicate, 10-20 cm long, densely pubescent, the verticills 10-20(or more)-flowered, subglobose, the bracts 3-4 mm long, hirsute and glandular; pedicels slender, hirsute, to 5 mm long. Calyx campanulate, 1.5-4 mm long, hirsute and glandular, the upper lip erect, broadly ovate oblong, the other teeth narrow, acute. Corolla pale blue or mauve to pink, 8-12 mm long, the tube declinate, 3-4 mm long, the upper lip to 4.5 mm long, 3 mm wide, erect, puberulent, the lower lop to 5-6 mm long, 4 mm wide, concave. Stamens with filaments mostly fused into a tube around the style. Fruit nutlets smooth, pale brown, ca. 0.7 mm long, 0.5 mm wide.

USES:

Plant juice is taken with honey for 3-6 days in case of morning rhinitis and cold in children. Leaf juice is used for cold, cough and fever in children. Crushed leaf juice is applied over the forehead for cold and running nose in children.

Plectranthus barbatus Andr. Lamiaceae

Indian Coleus, Patharchur, Marundu Koorkan

It is a perennial plant that grows to about 45-60 cm tall. It has four angled stems that are branched and nodes are often hairy. Leaves are 7.5 to 12.5 cm in length and 3 to 5 cm in width, usually pubescent, narrowed into petioles. Inflorescence is raceme, 15 – 30 cm in length; flowers are stout, 2 to 2.5 cm in size, usually perfect and calyx hairy inside. Upper lip of calyx is broadly ovate. The blue or lilac corolla is bilabiate. Lower lobes are elongated and concave so that they enclose the essential organs. The ovary is four parted and stigma is two lobed.

USES:

The roots are used as a potential drug for hypertension, congestive heart failure, asthma, bronchitis, cough and digestive comfort.

Pluchea lanceolata C. B. Clarke Asteraceae

Rasana, Phaar, Arattai

An undershrub, 30-60cm high, stem and branches terete, ashy and pubescent. Leaves: sessile, very coriaceous, 25-27 × 0.6-1.6cm, oblong or lanceolate, obtuse apiculate arranged at the base, finely ashy, pubescent on both sides. Flowers: capitula terminal; stalk upto 0.2 cm long; involucre bracts numerous, outer ones oblong, pubescent, apex acute, margins ciliate, tips often divided into 1-2 shallow lobes; inner ones linear, accute; ray- florets 0.4 cm long, fertile, pappus plumose, style slender, exserted; disc-florets few, 0.5 cm long, pappus 0.4 cm long, connate at base; corolla white-creamish or pale-yellow; anthers appendiculate at apex, style 0.4 cm long, 2-fid. Fruits: cypsels oblong, 0.05cm long, smooth.

USES:

The Whole plant is used as classical drug for arthritis. Its decoction is given for rheumatic conditions, muscular pains, edema, and fever and also applied externally as massage oil. The leaves used as a laxative, analgesic and antipyretic.

Plumbago zeylanica L. Plumbaginaceae

White Leadwort, Chitrak, Kodiveli.

Erect subshrubs, sometimes subscandent; stem striate. Leaves 3-7 × 1.7-3.2 cm, ovate or elliptic-ovate, base shortly and abruptly attenuated into a short petiole, apex acute or subacute, somewhat glaucous beneath; petiole ca. 1 cm long, amplexicaul at base and often auriculate. Flowers white, in elongate, terminal, often panicled racemes; rachis with sessile glands; bracteoles ovate, acuminate. Calyx 8-12 mm long, narrowly tubular, densely covered with stalked glands; lobes small. Corolla 1-1.6 cm across; tube ca. 1.5 cm long; lobes obovate-oblong. Stamens 5; filaments ca. 1.5 cm long. Ovary ellipsoid, 1- locular; stigma 5-fid. Capsule oblong, acute with 5 furrows.

USES:

It is advised to wear its flowers on the head for quick delivery. Crushed root mixed with sour buttermilk is given for two days to induce abortion. It is used to alleviate symptoms of bronchitis, cough and asthma. Root powder is used to rheumatoid arthritis, joint pains and muscular pains.

Pongamia pinnata (L.) Pierre Fabaceae

Indian beech

Medium-sized trees, branches spreading, twigs glabrous. Leaf imparipinnate, rachis 5-15 cm long, leaflets 5-9, c. 5-10 cm long, elliptic or ovate-oblong, acute-acuminate, glabrous; petiolule c. 5-8.0 mm long; stipels absent; stipules small, caducous. Raceme shorter than the leaf. Pedicel c. 7-10 mm long. Bract c. 2.5 mm long. Calyx c. 3.5-4.0 mm long, campanulate, pubescent, obscurely toothed. Corolla white tinged with violet or pinkish. Vexillum c. 1.2-1.3 cm long. Stamens 10, monadelphous. Fruit c. 3.5-5.0 cm long, c. 1.7-2.5 cm long, obliquely oblong, woody, indehiscent, usually 1-seeded.

USES:

This plant is having anti-inflammatory, antimicrobial, and anti-diabetic properties. All plant parts are used in treating skin infections, ulcers, hemorrhoids, and rheumatism.

Premna mollissima Roth Verbenaceae

Head Ache Tree, Munnai

Trees, to 7 m high. leaf is simple, bright green-coloured above and paler beneath. Leaf coriaceous, glossy and glabrous on both sides. Petiole 1-1.2 cm long with a pair of ensiform stipules at the axil of the stalk. Lamina 6-8 cm long × 3-4 cm wide. obovate or elliptic, the apex is acute-acuminate, base is cuneate or rounded and the margins are entire or finely serrate. venation of the leaf is reticulate with a prominent midrib. Flowers greenish-white, in terminal corymbose panicled cymes; bracts small. Calyx small campanulate, 2 lipped, 5 lobed. Corolla tube short, villous inside, lobes 5. Stamens 4, didynamous, inserted below the throat of the corolla tube; anther ovate. Ovary superior, 2-4-celled, ovules 4; style linear; stigma shortly bifid. Fruit a drupe, seated on the calyx, globose, purple; seeds oblong.

USES:

The root is anti-inflammatory, stomachic, anti-pyretic. Prescribed for liver complaints, colds, obstinate fevers, flatulence, urticaria, rheumatism, neuralgia. Roots part of Ayurvedic medicine "dashamula" used for various ailments.

Prunus americanus

Marsh Rosaceae

A large shrub or small tree, up to 4.5 m. The branches are thorny. The leaves are alternately arranged, with an oval shape. The leaflength is usually 5.- 10. cm long. The upper surface of the leaf is dark green; the underside is smooth and pale. The small white flowers with five petals occur singly or in clusters in the leaf axils. The globular fruits are about 2.5 cm in across.

USES:

A tea made from the scraped inner bark is used as a wash to treat various skin problems and as a mouth wash to treat sores. A poultice of the inner bark is disinfectant and is used as a treatment on cuts and wounds. The bark is astringent, diuretic and pectoral. It has been used to make a cough syrup. An infusion has been used in the treatment of diarrhoea, kidney and bladder complaints. An infusion of the twigs has been used in the treatment ofasthma.

Pseudarthria viscida (L.) Wight & Arn. Fabaceae

Sticky Desmodium, Chapakna, Movilai

Viscid pubescent subshrub up to 75 cm tall. Leaves 3-foliolate; terminal leaflet 5-8 × 4-6 cm, ovate-rhomboid; laterals 3-4 × 2-3.5 cm, obliquely ovate, apex acute, base cuneate or obtuse, glabrescent above and densely woolly below; petioles 4-6 cm long; stipules 5-6 mm long, subulate. Flowers in axillary or terminal, often branched racemes; bracts and bracteoles 4-5 mm long, subulate, hairy. Calyx-tube 1-1.5 mm long; lobes 2-3 mm long, subulate, hairy. Corolla purplish; standard 4-5 mm long, ovate-obtuse. Stamens 9+ 1; staminal sheath c. 4 mm long. Ovary 3.5-4 mm long, terete, densely pubescent without; style subulate, incurved; stigma capitate. Pods 1-2 × 0.4-0.5 cm, linear-oblong, compressed, pubescent with hooked hairs; seeds 4-6, reniform.

USES:

Roots are used to treatment ofrheumatic arthritis, diarrhoea, asthma, piles, dysentery, and to cure wounds. It serves as a cardiotonic. It is also helpful in reducing blood sugar levels.

Psidium guajava L. Myrtaceae

Guava, Amrood., Koyya

Small trees. Stem smooth with pealing bark. Young stem 4-angled. Leaves 16-11 × 2.5-5 cm, elliptic-oblong, base rounded to obtuse-cuneate, apex acute-apiculate, hirsute on both sides when young, glabrous on ageing except the nerves, thin-coriaceous, lateral nerves prominent; petioles 0.6-1 cm long. Cymes axillary, 1-3-flowered; peduncles 0.5-1.2 cm long; pedicel short 0. Calyx tube 4-9 mm long, ovoid, densely hirsute; lobes 4, unitedland closed in bud. Petals 4, white, 1-1.5 cm long, broadly ovate, caducous. Stamens many. Ovary globose, many-celled; ovules numerous; style subulate. Berry 2.5-3.5 cm diam., globose crowned by persistent calyx lobes. Seeds many, embedded in fleshy pulp.

USES:

It is used in the treatment of gastrointestinal diseases such as vomiting and simple diarrhea to the treatment of wounds, caries and cough.

Psoralea corylifolia L. Fabaceae

Purple Fleabane, Babchi, Karbooga arisi

It is an erect annual herb, 60 cm high. Stems and branches covered with conspicuous glands and white hairs. Leaves broadly elliptic, rounded and mucronate at apex. Inflorescence a raceme, axillary, comprises 10-30 flowers with a hairy pedicel. The seed oblong, flattened, rough, dark-brown having an agreeable aromatic odour. The taste is aromatic and bitterish.

USES:

It is used in treating skin disorders like vitiligo, psoriasis, and leprosy, as well as for curing digestive and canliovascular issues.

Pterocarpus marsupium Roxb. Fabaceae

Indian Kino tree, Vijaysar, Vengai

A tall. deciduous trees, to 25 m high. Stem exudation blood-red. Leaves imparipinnate, alternate; stipules small, lateral, cauducous; rachis 6.5-11.1 cm long, slender, pulvinate, glabrous; leaflets 5-7, alternate, lamina 3.5-12.5 × 2-7 cm, ellipticoblong, oblong-ovate or oblong, base obtuse or acute, apex obtuse and emarginate, margin entire, glabrous, coriaceous; Flowers yellow, in terminal and axillary panicles; 10-12 mm long; bracts small, dioecious; bracteoles 2, cauducous. Calyx-tube campanulate, lobes short, the upper 2 often connate. Corolla exserted; petals 5, all long-clawed, crisped along the margins; standard orbicular, wings oblique, obovate, auricled; keel petals oblique, small, slightly connate. Stamens 10, monadelphous; filaments subequal; anthers uniform; ovary shortly stalked, style filiform in curved, stigma capitate. Fruit a pod, 2.5-5 cm across, orbicular-reniform, broadly winged; seed one, subreniform.

USES:

The gum is used as an Astringent, anti-diarroheal and anti-haemrrhagic. Leaves are used externally to treat boils, sores, and other skin diseases. The gum exude 'kino', derived from this tree, is used as an astringent. The gum is bitter with a bad taste. However, it is antipyretic, anthelmintic and tonic to liver, useful in all diseases of body and styptic vulnerant and good for griping and biliousness, opthalmiya, boils and urinary discharges. The flowers are bitter, improve the appetite and cause flatulence.

Pterocarpus sant alinus L. Fabaceae

Red Sandalwood, Lalchandan, Semmaram

It is a small to medium sized deciduous tree, with an extremely hard, dark purple heartwood with a bitter flavour. Leaves are 3 foliated, 10-10 cm long and rachis swollen at base. Generally, there are 3 leaflets, broadly egg-shaped or orbicular. Base is round or slightly heart-shaped. Apex is also rounded or deeply notched. Margin is entire, leathery, shiny, hairless and distinctly stalked. Flowers are bisexual, stalked in auxiliary simple or sparingly branched racemes, yellow, about 2cm long, fragrant. Pods are unequally orbicular, flat about 5x4.5 cm including the wing, and gradually narrow into a short tip about 1cm long. Seeds 1 or rarely 2, more or less kidney shaped, 1-1.5cm long, smooth, reddish brown

USES:

The bark helps in lowering blood sugar levels by enhancing insulin secretion. It acts as a detoxifier and aids in reducing blood toxins. It promotes healing of wounds due to its healing properties. It helps in managing diarrhea and dysentery. It is also used in the treatment of inflamed joints and headaches.

Punica granatum L. Punicacae

Pomegranate, Anar, Madhulai

It is a small multi-stemmed a small tree up to 5m tall. Canopy open, crown base low. Stem woody and spiny, bark smooth and dark grey. Leaves simple, 2-8 cm long, oblong or obovate, glabrous, oppositely placed, short-petioled surface shining. Flowers regular, solitary or in fascicles at apices, 4-6 cm. Petals lanceolate, 5-7, wrinkled and brilliant orange-red. Hypanthium coloured, 5-8 lobed. Anthers numerous. Calyx persistent. Fruit a round berry, 5-12 cm, pericarp leathery. Interior compartmentalized with many pink-red sections of pulp-like tissue, each contains a seed grain. Fruits globose, 12 cm across, with reddish thick husk and a coriaceous woody rind. Seeds numerous, angular with fleshy testa, 1.3 cm long.

USES:

The astringent features of the flower juice, rind and tree bark are considered valuable for a wide range of purposes, such as stopping nose bleeds and gum bleeds, toning skin (after mixing with mustard oil) firming-up sagging breasts and treating hemorrhoid. It may be also used to improve sperm count and semen quality.

Quercus infectoria G. Oliver Fagaceae

Oak Gall, Majuphal, Masikkai

It is tree up to 5 m tall with many spreading branches. The bark is slightly grey in colour. The leaves 10 cm long, ovate to oblong mucronate, rounded, smooth, unequal at the base and shiny on the upper side. The galls arise on young branches of the tree, when gall wasps sting the tree and deposit their larvae, the chemical reaction causes an abnormality in the oak tree causing hard balls. The galls are globular in shape from 10 to 25 mm in diameter. Th galls are hard and heavy usually sinking in water.

USES:

The gall is useful for its astringent action. It helps in controlling diarrhoea, leucorrhoea, uterine prolapse, rectal prolapse, irritable bowel syndrome, and for tightening the loose or sagging tissues.

Rauvolfia serpentina L. Apocynaceae

Indian snakeroot, Sarpagandha, Sarpagandhi

It is an erect perennial shrub, that reaches about 60 to 90 cm in height. The leaves are borne in whorls of three to four, 8 to 18 cm long and 4 to 6 cm broad, lanceolate, acute, glabrous, bright green on the upper and pale on the underside. The flowers are borne in a two- to four-nate cyme. The calyx is five lobed and the corolla is salver-shaped, its tube pinkish, cylindrical, slender. In early stages, the inflorescence is light green, but when the flowers open, they show white corollas with a reddish tinge on the outer side. At a later stage, in the mature inflorescence, the axis, pedicels and calyx assume a bright coral red colour. The spherical fruit is a drupe 0.5 cm in diameter, single or double, the colour changing from green to red and finally to dark blue. The roots are cylindrical and tortuous or curved. The tap root of a two-year old plant usually reaches a length of30 to 50 cm and a diameter of 12 to 15 mm.

USES:

Whole plant extract is used for insanity, hypertension and psychotic disorders like schizophrenia, anxiety, epilepsy and insomnia. Root ground in rice cooked water is heated and applied for burning sensation and uneasiness in body. Root extract with lime juice is taken for snake bite. Root decoction is used for skin diseases.

Rhaphidophora pertusa (Roxb.) Schoot Araceae

Perforated Philodendron, Elitthadi, Yanaithippli

Stout perennial climber, with stem cylindrical, 4 cm thick, fleshy. Leaves are 25-35 × 25-30 cm, broadly ovate, pinnately cut or rarely entire, sometimes perforated with large holes, tip pointed, base rounded; leaf-stalk 20-35 cm long, channelled. Flowers are borne in solitary florescence in leaf-axils; spathe up to 20 × 8 cm long, ovate-oblong, concave, pointed, creamy white; spadix 12 cm long, 2 cm thick. Flowers are unisexual, densely packed. Female flowers are basal; ovary obconical, flat, I-celled; ovule solitary; stigma knob-like. Male flowers are above; stamens 4-6, free.

USES:

The stem is used to treat pain in the colon and abdominal tumors. The leaves are used to treat inflammatory conditions. The juice of the plant with black pepper is used as an antidote against snake bites.

Rheum australe D. Don Polygonaceae

Himalayan Rhubarb, Archa

It is a tall (1.5-3 m), robust and leafy perennial herb. The stem is glabrous or pubescent, streaked green and brown with purple to red shade. Rhizomes are 15-30 cm long with a dull orange to yellowish brown surface The leaves are thick, dull green, highly wrinkled with distinctly rough surface, orbicular or broadly ovate, cordate based on 5-7 nerves, sub scaberulous above and papillose beneath, entire margin and sinuolate with an obtuse apex plant has dark reddish-purple flowers in densely branched clusters in a long panicle inflorescence which can be 30 cm long The inflorescence is fastigiate branched and densely papilliferous which greatly enlarges in fruit. The flowers are small, 3 mm in diameter, in axillary panicles The perianth is spreading, 3-3.5 mm in diameter with outer parts smaller and oblong-elliptic Fruit is ovoid-oblong, ovoid-ellipsoid or broadly ellipsoid, 13 mm long and purple in color.

USES:

The root is to cure a wide range of ailments related to the circulatory, digestive, endocrine, respiratory and skeletal systems as well as to treat various infectious diseases.

Rhinacanthus nasutus (L.) Kuntze Acanthaceae

Snake Jasmine, Palak-juhi, Nagamalli

Erect shrubs up to 1.5 m, stem terete, tomentose. Leaves up to12 × 6 cm, elliptic, acute at both ends, crenulate, minutely pubescent, nerves 7-10 pairs.. Flowers sessile in terminal cymes; bracts and bracteoles similar, 2 mm long, hispid; sepals 5, 5 mm long, linear-lanceolate, hispid; corolla white, tube 25 mm long, slender, hispid; upper lip entire, oblong, acuminate; lower lip broad, 3-lobed, obtuse; stamens 2. style slender. Capsule 2 cm long, clavate, with a lower solid slender stalk, glabrous; seeds 1 or 2 in each cell, rugose.

USES:

The root is used as an antidote against snake venom. The leaves can be drunk as a tea or made into a balm and is also traditionally used to treat a range of disorders including ringworm and inflammatory disorders.

Ricinus communis L. Euphorbiaceae

Castor bean, Arandi, Aamanakku.

Shrub, up to 2.5 m high; branchlets glaucous, glabrous. Leaves alternate, peltate, palmately 7-8-lobed; lobe oblong-lanceolate, 7.5-13 × 3-7, cordate at base, serrate at margin, acuminate at apex, glabrous; 8-palmalety nerved, 2-gland at base; petiole 22 cm long. Panicles elongate, terminal, up to 25 cm long. Flowers ca. 1 cm across, greenish. Male flowers lower elongate. Femsle flowers upper erect. Calyx-5, greenish. Petals 0. stamens many; filament branching; anther ovoid, distinct. Ovary globose, 3-celled, 04 mm across; 3-loculed; ovules 3; style 2-fid, linear, 03 mm long; stigma papillose. Capsules globose, 1.4 cm across, softly prickly, 3-lobed; seeds 3, oblong, 1 cm long, brownish.

USES:

Root decoction is used as a purgative in rheumatic conditions and constipation. Massage with seed oil is given for ankle twist. Tender shoot tip boiled in milk or rice washed water is given for jaundice in pregnant women. Leaf(smeared with castor oil) is heated and tied overnight the head for dandruff. Oil prepared by grinding its root, turmeric and pepper (4:2:1) powder with gingelly oil and keeping it in sun light is applied for all types of wounds and cuts.

Rivea hypocrateriformis (Desr.) Choisy Convolvulaceae

Midnapore Creeper, Misttai

Twining subshrubs, stem white silky-pubescent. Leaves 3-4 × 3-6 cm, cordiform, apex obtuse, base cordate, subcoriaceous; petiole to 6 cm. Flowers axillary, solitary or in cymes; bracts lanceolate. Crolla white, 7 cm across, salver-form. Stamens 5, unequal, to 1.8 cm, anthers oblong, 4 mm. Ovary oblong, 4 mm, 4-locular, each I-ovulate, style 3.5 cm. Capsule conical, indehiscent, 1.5 × 1 cm, I-celled; seeds 4 or a few.

USES:

One of the main uses of this plant in Ayurvedic medicine is as a contraceptive. When taken in the very early stages, it can prevent implantation of the foetus and is known to completely interrupt early pregnancy and influence the estrous cycle. Also it is used in treating cough, headache, skin disease as well as treating external conditions such as burns, piles and to relieve pain.

Rosa × damascena Mill. Rosaceae

Damask Rose, Gulab, Roja

It is a perennial bushy shrub reaching approximately 1 to 2 meters in height. The stems are densely armed with stout curved prickles, weakly compressed, subulate, usually mixed witth stalked glands and setae. The leaves are imparipinnate and compound with 5-7 leaflets, leaf-stalks prickly.. Flowers 3-5 or more, pink, rarely white, double, usually smelling. Pedicels rather long, covered with stalked glands. Hypanthium usually covered with stalked glands. Fruit globose or ovoid, often undeveloped. The roses are a light to moderate pink to light red. The relatively small flowers grow in groups.

USES:

The flowers are used for treatment of abdominal and chest pains, strengthening the heart, menstrual bleeding, digestive problems and constipation. Also helpful in reducing the inflammation, especially of the neck and to treat irregular menstruation.

Rosmarinus officinalis L. Lamiaceae

Rosemary

It is an evergreen, usually erect, bushy shrub up to 2 m tall and wide. Stem indistinctly quadrangular, finely grey pubescent. Leaves opposite, tufted on the branches, sessile to short petiolate; blade linear, 1-5 cm × 1-2 mm, base attenuate, margin entire but revolute, apex obtuse, leathery, aromatically fragrant when crushed. Inflorescence racemose, axillary, 5- to IO-flowered. Calyx campanulate, 2-lipped, 5-6 mm long. Corolla tubular, 2-lipped, 10-13 mm long, pale blue or blue. Two anterior stamens perfect, 7-8 mm long, ascending under the base of the upper lip, 2 posterior stamens reduced to hardly visible staminodes. Pistil with deeply 4-partite ovary, style incurved, 1.5 cm long ending into 2 short, unequal branches with stigma. Fruit composed of 4 subglobose to obovoid nutlets, about 2 mm long, glabrous and smooth.

USES:

It is traditionally used to help alleviate muscle pain, improve memory, boost the immune and circulatory system, and promote hair growth.

Rotheca serrata (L.) Steane & Mabb. Acanthaceae

Blue Fountain Bush, Barangi, Sirutekku

A shrub about 1.5 m tall. Branches obtusely quadrangular. Leaves simple, opposite or whorled in 3's, lanceolate, ovate-oblanceolate, 6-34 × 2-10 cm across, base cuneate, margins serrate, apex shallow acuminate or acute, chartaceous, green, slightly pubescent above and slightly paler slightly pubescent beneath. Flowers bisexual, zygomorphic, pedicel reddish, about 0.3-0.5 cm long, pubescent in terminal thyrses, 30 cm long. Petals oblong to obovate, 1.2 cm long. Stamens 2-4 cm long, protruding out. Style protruding out. Fruit drupaceous, glabrous, succulent, about 1 cm in diameter.

USES:

Root is used in treating Asthma, bronchitis, chronic cough and it is given in reducing joint inflammation and trating gout. Apoultice of leaves is used to treat skin diseases, infections and herpes. It has been used for managing diabetes, cancer and as an antidote for snakebites. Leaves are vermifuge and bitter tonic.

Rubia cordifolia L. Rubiaceae

Indian Madder, Manjitha, Manjatti

Perennial climbing herb. Stem divaricately branched, branches, rigid, quadrangular, glabrous or prickly-hispid, specially on the angles. Leaves (2-) 4 (-8) in a whorl, one pair of each whorl usually larger than the other with longer petioles, 1-10 × 2.1-6 cm, ovate-cordate, entire, acute, acuminate-cuspidate, cordate at base rarely rounded, 3-9 palmately veined, upper surface mostly glabrous and rough. Flowers 3.5-4.5 mm in diameter, (4-) 5-merous, pedicel 3.4-5 mm, quadrangular. Calyx c. 1 mm long, globose, glabrous. Corolla greenish-white, tube± absent, lobes (4-) 5, lanceolate, acute, c. 3-3.5 × 1-1.5 mm. Stamens 5, 0.5-0.75 mm, anthers oblong, basifixed. Styles 2, united at the base, c.0.6 mm long; stigma globose. Fruit 4-5 × 3.5-5 mm, didymous or globose by suppression of one carpel, purplish black when ripe.

USES:

It is a blood purifying herb. It cools and detoxifies the blood, removes stagnant blood and dissolves obstructions in the blood flow. And also helps in regulating menstrual flow and manage menstrual pain. It can also be used to treat for kidney and bladder stones, managing fever and as a natural dye.

Ruta graveolens L. Rutaceae

Garden Rue, Sadab, Aruta

It is a perennial shrub up to 1 m tall. Leaves pinnately divided, bluish- green up to 12 cm long, have oblong/ spatulate segments. It has a pungent aroma when bruised and leaves have a bitter taste. Flowers yellow, small, 4- to 5-petaled, dull yellow in flattened corymbs. Petals are with more or less undulate or denticulate margins Fruit is a capsule with brown seed.

USES:

This plant is used for its anti-inflammatory, antispasmodic and anti-arthritic properties, often treating joint pain, sciatica and muscle spasms. It acts as an emmenagogue to regulate menstruation and helps with digestive issues. The juice of the plant has been used in treating nervous system disorders, skin issues and serves as an in secticide.

Sacchararum officinarum L. Poaceae

Sugarcane, Ganna, Karumppu

It is a perennial giant tropical grass growing erect and tall. It is known for its unique growth pattern, where lateral shoots emerge from the base, leading to the development of multiple stems. These stems typically grow to a height of3-4 meters. The plants are with stout, jointed, long, narrow leaves that grow from the base. The terminal inflorescence is a panicle, characterized by a pinkish plume that is broadest at the base and tapers towards the top. The spikelets are borne on side branches and are concealed in tufts oflong, silky hair. The fruits are dry and contain a single seed, fruit is a caryopsis.

USES:

It is used in treating diabetes, heart health and seives as adiuretic. It's also used to treat diabetes, myocardial infarction and manage cholesterol levels. hemorrhage.

Salacia chienensis L. Hippocrateaceae

Chinese Salacia, Saptarangi, Karukkuvai

It is a small, erect or straggling tree or climbing, smooth shrub, reaching a height of 5 meters or more; branchlets angular. Petiole 8-10 mm; leaf blade elliptic, narrowly ovate-rotund or obovate-elliptic, 5-11 × 2-5 cm, leathery, abaxially brown-green when dry, adaxially shiny olive-green, base cuneate, margin sparsely denticulate, apex subacute to shortly acuminate; lateral veins 6 or 7 pairs. Fascicles axillary, 3-6-flowered; pedicel 6-10 mm. Sepals deltoid, up to 1 mm, margin ciliolate; petals broadly ovate, ca. 3 mm, broadly spreading or curved outward, apex rounded. Disk cupular, ca. 1 mm tall. Stamens 3; filaments short, flat, inserted at margin of disk; anthers forked. Ovary concealed in disk, 3-loculed; ovules 2 per locule; style very short, conic. Berry globose or ovate, ca. 2 cm in diam., red when ripe, I-seeded.

USES:

In addition to treating diabetes, it is also widely used for treating gonorrhea, asthma, itchiness, joint inflammation, obesity and menstrual disorders.

Sansevieria roxburghiana Schult. & Schult.f. Agavaceae

Indian Bow String Hemp, Marul. Manji

Rhizomatous herbs, with creeping rhizome. Leaves in a tuft, broad towards the middle, rigid, cross-striped, with rigid spine-like tip flowers greenish-white tinged with violet, in fascicles of 3-6 on long racemes. Perianth tubular; lobes 6, narrow. Stamens 6, curved and exerted. Fruits 1-3 seeded berry.

USES:

Leaf juice is given along with honey for cough, bronchitis and chest pain. Rhizome decoction is used in treating chronic persistent coughs, colds, asthma and for reducing sugar levels. Leaf paste is applied for poisonous bites. Heated leaf juice is poured into the ear for earache.

Santalum album L. Santalaceae

Indian Sandalwood, Chandan, Chandanam

Tall trees grows between 4 and 9 metres in height, though it can reach up to 20 metres in ideal conditions. It often has slender, drooping branches and a high bushy crown. In young trees, the bark is smooth and grey-brown; in mature trees, it becomes dark brown to nearly black with deep vertical cracks. The leaves are simple, opposite, and thin. They are typically ovate or elliptical, measuring 3-8 cm long and 3-5 cm wide, with a shiny green upper surface and a paler, glaucous underside. Flowers small and unscented, the flowers are initially straw-coloured or green, turning purplish-brown, violet, or red as they mature. They grow in small clusters known as paniculate cymes. Fruit a fleshy, globose drupe that is purple to black when ripe, approximately 1 cm in diameter, containing a single seed.

USES:

Hot wood paste is applied all over the body of children suffering from fever and is taken internally for skin diseases. Heart wood extract with tender coconut water is used as a blood purifier for urinary tract infection. Heart wood extract with tender coconut water is given to drink in case of rashes and allergy. Sandalwood and red sandalwood paste is applied (after heating) over the forehead for headache in children.

Saraca asoca (Roxb) de Wilde Caesapiniaceae

Asok tree, Ashoka, Asogam

It is a small evergreen tree 5-7 cm high. Trunk often tubercled; bark lenticellate, dark, shallowly fissured; blaze purplish. Branchlets terete, glabrous. Leaves compound, paripinnate, alternate, distichous; stipules caducous; rachis pulvinate, 7-30 cm long; petiolule 0.1-0.6 cm long; leaflets opposite, 4-6 pairs, lamina 6-31 × 1.5-9 cm, narrow elliptic-oblong or lanceolate, apex acute to acuminate, base acute to rounded or subcordate, subcoriaceous, glabrous; midrib raised above; secondary nerves ca. 11 pairs, looped; tertiary nerves reticulate. Inflorescence dense corymbs. Flowers are polygamous apetalous, yellowish orange turning to scarlet, in short laterally placed corymbose, axillary panicles, bract small, deciduous. Calyx petaloid; fragrant. Pod, flat, oblong, to 15 × 4.5 cm, apiculate; Seeds are 4-8, ellipsoid-oblong and compressed.

USES:

It is used to regulates menstrual cycles, treats uterine disorders and supports reproductive health. The bark is used arthritis, reducing joint pain. Traditionally piles, indigestion, constipation and diarrhea.

Sarcostemma viminale (L.) R. Br. Asclepiadaceae

Moon creeper, Soma, Somavalli

Stem cylindrical, 3-7 mm in diameter, green, glabrous, intemodes 10-20 cm long. Flowers sweet-scented in sessile many-flowered terminal umbels. Pedicels 6-12 mm long, pubescent. Bracts minute, linear-lanceolate. Sepals 5, 1-1.5 mm long, subacute, margin membranous, glandular inside near the base. Corolla campanulate, pale greenish white, divided almost to the base, lobes c.5 mm long, ovate-oblong, subacute. Outer corona crenately IO-lobed, lobes of the inner corona thick, fleshy obtuse, longer than the staminal column and concealing the anther. Staminal column very short. Pollen-masses compressed, clavate. Follicles 10-13 cm × 8 mm, lanceolate, tapering at both ends, slightly diverging when 2 together. Seeds 3-6 mm long, flattened ovate. Coma 1.8 cm long.

USES:

The plant has been reportedly used to treat a wide variety of conditions including cuts wounds, warts and skin infections. Used as a laxative to treat diarrohoea and constipation. Also applied as a paste for fractured bones.

Satyrium nepalensis D. Don Orchidaceae

Nepal Satyrium, Salam Misri, Bullock's Homs

It is terrestrial herb 20-70 cm tall. Stem 1-3 membranous sheaths at base, 1- to 3- leaved Tubers oblong-ellipsoid to ellipsoid, 2-5x 0.5-1.5 cm. Leaves basal and subopposite, or sometimes cauline and widely spaced and alternate, broadly ovate, ovate-lanceolate, or lanceolate-oblong, slightly fleshy, margin somewhat crisped, apex acute or acuminate Flower fragrant, widely spreading, whitish, pink, or pale purple, glabrous. Calyx d, orsal sepal narrowly oblong-elliptic, 4-6 × 1-1.8 mm, apex obtuse; lateral sepals oblong to subovate, slightly oblique, 4-6 × 1.5-2 mm, apex obtuse. Corolla narrowly oblong to narrowly elliptic, 3.5-5 × 1-1.2 mm, outer surface carinate, margin sometimes finely ciliate, apex acute to obtuse and sometimes incised; lip hooded, subglobose, 5-6 × 4-6 mm, outer surface carinate, 2-spurred, margin sometimes irregularly dentate, apex acute or obtuse and often deflexed; spurs parallel, slender and cylindric, to stout and conic, curving downward, 3-13 mm, or reduced and saclike, or rarely absent altogether Fruit capsule, spindle shaped. Seed microscopic, powdery.

USES:

It has been used in ethnomedical studies to cure malaria, diarrhea, and dysentery since ancient times. Because of its varied chemical makeup, which contains triterpenes, alkaloids, flavonoids, and unsaturated sterols, it is recognized as an orchid with therapeutic benefits.

Saussurea costus (Falc.) Lipsch. [Saussurea lappa (Decne.) Sch. Bip.] Asteraceae

Costus Root, Pachak, Goshtam

Herbs, 1-2.5 m high, pubescent. Roots stout, inverted cone-shaped, up to 60 cm long, greyish to dull brownish, with characteristic odour. Basal leaves scaberulous above, glabrate beneath, margins irregularly toothed; petiole 60-90 cm long, lobately winged; cauline leaves serni-amplexicaul, 15-30 cm long, sessile or with shorter petiole. Involucral bracts ovate-lanceolate, glabrous, purple. Flowers in heads, 2.5-4 cm in diam., axillary or terminal clusters of2-5, sessile. Petals dark purple. Anther tail fimbriate. Achenes ca 8 mm long, compressed, curved. Pappus hairs brown, double feathery.

USES:

The root of this plant act as a bronchodilator to alleviate asthma, bronchitis and coughs. It helps in reducing stomach aches, flatulence and diarrhea. It is also used in joint pain, rheumatoid arthritis and chronic pain management. It is potential in protecting liver cells and managing urinary stones.

Scoparia dulcis L. Scrophulariaceae

Sweet Broom Weed, Ghoda Tulsi, Kallurukki

It is a subwoody erect plant 50 cm tall with polygonal stem. Leaves are opposite or verticillate by three: blade is oblanceolate, 2.5-5 cm long and 1.5 cm wide, base is attenuated in acute comer, forming a pseudo-petiole, glabrous on both sides and riddled with bright green glands, margin is entire in the lower half of the limb and toothed in the upper half. Flowers are solitary or in pairs in the axils of the leaves, white. Calyx is formed of 5 free sepals almost to the base, elliptical in shape, ending in a point at the top, pubescent. Corolla is formed of 4 petals, rarely 5, almost free to the base, oval, apiculated apex. Calyx and corolla is 3 to 4 mm long. Four stamens have anthers with 2 equal compartments. Ovary is surmounted by a filiform style not exceeding the corolla. The fruit is a dehiscent capsule, ovoid, surmounted by style. It is 4 mm long and contains many seeds.

USES:

This plant is used in the treatment of diabetes, high blood pressure and stomach issues. The leaves are used for curing bronchitis, coughs and fever.

Semecarpus anacardium L.f. Anacadiaceae

Marking nut Tree, Bhilawa, Chaeran Kottai

Trees, to 12 m high, bark 5-6 mm thick, grey to greyish-brown with tessellate scales, furrows longitudinal and horizontal; blaze dull red; exudation acrid. Leaves simple, alternate, crowded at end of branchlets, estipulate; petiole 20-45 mm long, stout, pubescent; lamina 15-50 × 10-15 cm, obovate-oblong, elliptic-oblong or obovate, base acute, obtuse, round or truncate, apex obtuse, round or emarginate, margin entire, glabrous above nerves 10-20 pairs. Flowers polygamous, greenish-yellow, 6-8 mm across, in terminal pubescent panicles. Calyx lobes 5, ovate, margin ciliate, imbricate, deciduous; petals 5, imbricate; disc broad, annular, villous. Stamens 5 inserted below the disc; filaments filiform, glabrous, in male flowers as long as or longer than petals, in female flowers shorter than petals; anthers oblong; ovary superior, densely villous, I-celled, ovule pendulous from a basal funicle; styles 3, divergent. Fruit a drupe, 2-5 cm long, fleshy, purplish-black, seated on an orange hypocarp as long as the drupe, pericarp rich in corrosive resinous juice; seed pendulous.

USES:

Ripe fruits acts as a neuro protective, antiinflammatory and showing promise in cancer treatment and boosting cognitive health. It is highly effective in treating rheumatoid arthritis. It is used for skin disorders like eczema and psoriasis and displaying immunomodulatory activity.

Senna alata L. Caesalpiniaceae

Ringworm Bush, Dadmardan, Seemai Agathi

A perennial shrub, up to 2.5 m high. Leaves pinnately compound, paripinnate; leaflets opposite, 10-12-pairs, brodaly oblong-obovate to oblanceolate, 5-9.5 × 2.3-5.5, oblique, truncate at base, entire, minutely ciliate at margin, retuse at apex, glabrescent; 9-13-pairs. Racemes terminal, up to 20 cm long. Flowers ca. 1.5 cm across, golden-yellow, subsessile. Calyx-lobe 5, obovate, 1.4 cm across, pale yellow. Petals 5, broadly obovate, 1.6 cm across, undulate at margin, truncate at apex, reticulate nerved. Stamens 1O; upper 3 staminides; filament 03-7.5 mm long, glabrous; anther 2-celled, 03-9.5 mm long; medium size filament 2.5 mm long; anther equal, 03 mm long; lower anther globose, 1.5 mm long. Ovary angled, 05 mm long, style slightly flat, 1.2 cm long; stigma small. Pod 2- broadly winged, up to 11 cm long; seeds many, obovoid, triangular, 05 mm across, transverse.

USES:

It is used as a laxative or purgative and in the treatment of skin problems. For laxative purposes usually a decoction of the leaves is drunk, and less often the flowers, roots or the stem are used. The leaves are also known to be used in the treatment of bronchitis and asthma.

Senna alata L. Caesalpiniaceae

Ringworm Bush, Dadmardan, Seemai Agathi

It is a subwoody erect plant 50 cm tall with polygonal stem. Leaves are opposite or verticillate by three: blade is oblanceolate, 2.5-5 cm long and 1.5 cm wide, base is attenuated in acute comer, forming a pseudo-petiole, glabrous on both sides and riddled with bright green glands, margin is entire in the lower half of the limb and toothed in the upper half Flowers are solitary or in pairs in the axils of the leaves, yellow. Calyx is formed of 5 free sepals almost to the base, elliptical in shape, ending in a point at the top, pubescent. Corolla is formed of 4 petals, rarely 5, almost free to the base, oval, apiculated apex. Calyx and corolla is 3 to 4 mm long. Four stamens have anthers with 2 equal compartments. Ovary is surmounted by a filiform style not exceeding the corolla. The fruit is a dehiscent capsule, ovoid, surmounted by style. It is 4 mm long and contains many seeds.

USES:

This plant is used in the treatment of digestive problems, pulmonary conditions, fever, skin disorders, hypertension and hemorrhoids. The leaves are chewed to treat cough; they first taste bitter and later sweet (like liquorice). They were formerly used in the treatment of diabetes. The leaves are macerated in warm water and drunk copiously when cooled in the treatment of feverish headaches.

Senna auriculata (L.) Roxb. Caesalpiniaceae

Tanner's Senna, Tarwar, Aavarai

Shrub up to 2 m tall. Leaves alternate, paripinnately compound; stipules large and leafy, broadly reniform, 7-22 mm wide, persistent; petiole 10-14 cm long; rachis provided with a gland between each pair ofleaflets; leaflets 6--13 pairs, oblong-elliptical to obovate elliptical, 10-35 mm × 5-12 mm, rounded and mucronate at apex, glabrous to pubescent. Inflorescence an axillary raceme, 2-8-flowered. Flowers zygomorphic, 5-merous. Sepals rounded at apex. Petals free, unequal, 1.5-3 cm long, yellow. Stamens 10, the 3 lower ones largest and fertile, others usually sterile. Ovary superior, falcate, c. 1.5 cm long, stalked, style c. 1 cm long. Pod a flattened cylindrical 5-18 cm × 1-2 cm, transversely undulate between the 10-20 seeds, indehiscent. Seeds compressed ovoid-cylindrical, 7-9 mm × 4-5 mm, with a distinct areole on each face.

USES:

The flowers and leaves are used to control blood sugar levels. It is used to treat leprosy, skin diseases and di and for its wotmd healing properties. I is also usefil in the treatment ofuriruuy infections, excessive menstrual and gonorrhea. It is widely used for constipation, diarrhea and stomach iss ues.

Senna tora (L.) Roxb. (Cassia tora L.)Caesalpiniaceae

Sickle Senna, Chakunda, Thakara

A small weed, subshrub, up to 60 cm; branchlets glabrous. Leaves paripinnate, with subulate glands between the leaflets of the 2 lower pairs; leaflets obovate, glabrous above, pubescent beneath up to 4 cm long, 3 pairs, thin, coriaceous. Flowers terminal racemes, corymbose. Sepals 5, oblong to rounded. Petals 5, yellow. Perfect stamens 6-7. Pods linear, subtetragonous, puberulous, short, stipitate, flat, compressed. Seeds hard, 1 cm long, 3-4 mm thick, oblong or rhombohedral, both ends appear as if cut off obliquely, greenish-brown to brownish-black, smooth and shiny. Seed coat consists of longitudinally elongated cells, covered with thick, smooth cuticle.

USES:

The leaf paste is used to treat ringworm, itching, leprosy, psoriasis and scabies. It acts as a mild laxative to relieve constipation. The leaves and seeds are used to reduce inflammationin diseases like rheumatism, bronchitis and for general pain relief. Seeds are used for treating eye infections, improving vision and relieving glaucoma.

Sesbania grandiflora (L.) Poir. Fabaceae

Agastya's Plant, Hathya, Agathi

Small soft wooded tree up to 7.5 m tall. Leaves even-pinnate, leaves to 30 cm long. leaflets 25-30 pairs, oblong, 1-3.5 × 0.5-1 cm. Flowers 5cm across, cream in 3-5- flowered racemes upto 10 cm long. Calyx -tube campanulate, lobes triangular shorter than calyx tube. Corolla exceeding calyx; petals with long claws. standard reflexed, callose at base, wings falcate oblong; keels obtuse, incurved. Stamens diadelphous, vexillary stamen free, bent at the base with stamina sheath; anthers uniform. Ovary stipitate, ovules many, style incurved, glabrous. stigma capitates. Pod to 60 cm long, 0.5 cm across, margins thick, flattened, tardily dehiscent, septate between seeds. Seed transversely oblong, estrophiolate.

USES:

Flower juice is used for night blindness. Leaves and bark are used for crning dianhea, dysentery and stomach issues. It is widely used to treat rheumatism, gout, and arthritis related pain.

Side cordifolia L. Malvaceae

Heart leaf Sida, Bariyar, Sittamutti

It is an annual or perennial short, erect, greyish-green, softly hairy or pubescent woody undershrub, 0.5-1 m high. Leaves are simple, very downy, alternate, 2.5-5 cm×1.8-3 cm, orbicular, ovate, ovate-oblong or cordate, margin crenate, base cordate, petioled, stipulate, stipules linear. Flowers light or sulfur yellow to cream white, axillary and solitary but appears crowded in the upper part and toward tips of the branches, without an epicalyx. Fruits are depressed, globose schizocarp, 6-8 mm diameter, each carpel having two long straight linear to setaceous scabrous awns. Seeds are smooth, flattened, reniform, brown or black.

USES:

This plant is used to treat rheumatism, arthritis. neurological disorders, respiratory issues like asthma, bronchitis and sexual dysfuction. It is also known to promote weight loss and boost energy due to its ephedrine content.

Sida rhombifolia L. Malvaceae

Arrowleaf sida, Bala. Kurunthotti

An erect perennial undershrub 0.3 to 1.0 m tall. The leaves are much larger, up to 6 cm long and 2.5 cm large, rhomboid or elliptic with petioles 4-5 cm long and the lower surface of the lamina hairy. Inflorescence is an axillary, solitary flower on slender pedicel, 2 to 3 cm long, jointed above middle. Calyx segments triangular stipules needle-shaped, less than 8 mm long. Corolla of 5 pale yellow petals, 1.5 to 1.8 cm in diameter. The fruit is a dehiscent capsule with 8-12 carpels; the beak of each is carpel rather shortly bent, and not straight and pointed. Seeds are flattened, reniform, 1.25 to 2 mm long, dark brown.

USES:

Whole plant: juice-is mixed with water beneficial in spermatorrhoea; Seeds: useful in colic, tenesmus and gonorrhoea; Leaves: useful in nervous urinary diseases, cystitis, strangury, bleeding piles, haematuria, chronic dysentery, and leucorrhoea ; Root: juice is used for healing the wound; Bark: is ef effective in curing facial paralysis and sciatica. Also useful in opthalmia, rheumatism, shivering fits and helps in improving sexual strength.

Silybrium marianum (L.) Gaertn. Asteraceae

Milk Thistle, Doodh Patra, Pal Nerunjil

It is an annual to biennial herb, up to 1m high. Stem, glabrous or slightly downy, erect and branched in the upper part. The leaves are alternate, large, white veined, glabrous with strongly spiny margins. The inflorescence is large and round capitula, solitary at the apex of the stem or its branches, surrounded by thorny bracts. The florets are hermaphrodite, tubular in shape with a red-purple corolla. The fruits are hard skinned achenes 6 to 8 mm long, generally brownish in color with a white silk like pappus at the apex.

USES:

It is used to treat cirrhosis, jaundice, hepatitis, and gallbladder disorders. It is particularly used to treat the liver but can also be used to treat the kidney and gallbladder. It is often suggested as a treatment for alcoholic hepatitis and drug rehabilitation.

Simarouba glauca DC Simaroubaceae

Paradise tree, Lakshmi Tharu, Sourga Maram

It is an evergreen, small to medium-sized tree growing up to 15 min height, with a narrow crown, well-developed root system, and straight, cylindrical bole that can be at least 30 cm in diameter. The leaves are arranged alternately, are odd pinnately compound, up to 40 cm long, with 10-20 leaflets. The leaflets are up to 10 cm in length, dark green above, lighter below, with an entire margin and rounded leaf tip. Flowers are arranged in at branch-ends and in leaf-axils, in panicles. The calyx has 5 unfused, greenish sepals. The flower has 5 free yellowish-white overlapping petals. The fruit is an oval purple/black drupe at maturity.

USES:

Extracts from the leaves, bark, and other parts of the plant have been used to treat various cancers, including oral, breast, lung, cervix, rectum, and colon. It is also used in curing diabetes.

Solanum anguivi Lam. Solanaceae

Forest Bitterberry, Banbhatta, Karimulli

It is a much branched undershrub up tp 1.5 m tall. Stem is stout, often bluish, prickled. Branches covered with stellate hairs. Leaves 5-15 × 2.5-7.5 cm, ovate, acute at both ends, clothed with simple hairs. Flowers purple borne in racemose extra axillary cymes clothed with dark purple coloured hairs. Fruit is a berry, 0.8 cm across, spherical, dark yellow, when ripe. Seeds minutely pitted.

USES:

Plant used to treat cough, asthma, fever, dysuria, worms, nervous complaints and skin diseases. Fruits chewed as a remedy for coughs, chest pains and against high blood pressure; Fruits ground with pepper is applied on forehead to cure fever and cold.

Solanum pubescens Willd. Solanaceae

Thomless Turkey Berry, Jangli Bhata, Kattu Sundai.

It is an unarmed velvet-hairy shrub. Leaves are up to 12 × 7 cm, ovate-elliptic to triangular, tip pointed, base enequally flat, membranous, margins entire to wavy; leaf-stalk up to 8 cm, velvet-hairy. Flowers are borne in racemes in leaf-axils; flower-cluster-stalk to 5 cm, velvet-hairy; flower-stalks 2 cm, velvet-hairy. Flowers are blue; sepals 6 mm, lance-shaped, velvet-hairy; flower tube 5 mm, petals 6 mm, lance- shaped, velvet-hairy; ovary 1.5 mm, style 1 cm, stigma capitate. The fruits are around 1.2 cm diameter and is more bitter. Matured fruits are orange when ripe about 1 cm in diameter.

USES:

It has anti-inflammatory, anti-ulcer, and anticonvulsant properties.

Solanum virgianum L. S.xanthocarpum Schrad & H. Wendl

Wild Egg Plant, Kanteli, Kandankathiri

A very prickly diffuse bright-green herb, prickles yellow, sharp, straight, compressed. Leaves ovate or elliptic, acute at the apex, truncate at the base stellately hairy on both sides, prickly along the nerves, irregularly lobed, petiolate, petioles stellately hairy. Flowersin extra axillay few-flowered cyme, Pedicellate, pedicels stellately hairy, prickly. Calyx campanulate, densely hairy, prickly, lobes ovate-oblong. Corolla purple, lobes triangular, hairy outside. Berry globose, yellow. Seeds subreniform, minutely foveolate.

USES:

The whole plant is valued as an expectorant and diuretic. Root decoction is prescribed for chest pain and catarahh, One tea spoon of berry juice is taken to treat sore throat. Also used for cough, asthama, colic fever, loss of appetite, chest pains and tooth ache.

Spermacoce hispida L. Rubiaceae

Shaggu button Weed, Madanagranthi, Natthai choori

Erect or decumbent herbs up to 45 cm tall; stems hispid or glabrescent, thickened at nodes. Leaves 1-3 × 0.5-2 cm elliptic -ovate, obtuse or acute at apex, margins often incurved, rounded and narrowed at apex, hispid or pubescent on both sides, ciliate, glaucous beneath, subsessile; stipules interpetiolar, hispid. Flowers in axillary fascicles. Calyx-tube up to 0.3 cm long, elliptic-oblong, hispid; lobes 4, obovate. Corolla pinkish white tinged with purple, funnel-shaped, tube up to 0.5 cm long; lobes 4, oblong. Stamens4, between the lobes, exserted, versatile. Style filiform, up to 0.75 cm long, stigma capitate, recurved. capsules globose, 0.6 × 0.4 cm oblong, pubescent or hispid. Seeds up to 0.3 cm long, liner-oblong, plano-convex, grooved on the flattened face, finely granulate.

USES:

The whole plant ground with water or butter milk is applied for urticaria, rashes and scabies. Leaf juice is recommended for uterine bleeding. Seed decoction is a nervine tonic, also serve as a blood purifier, digestive system support and given for gall bladder stones. It is having potential anti-cancer activity.

Sphaeranthus indicus L. Asteraceae

East Indian Globe Thistle, Gorakhmundi, Kottaikkaranthai

It is a much branched, strongly-scented annual herb, up to 30 high; branchlets angled, glandular pubescent. Leaves alternate, oblong-oblonceolate, 5-6.5 × 2.2-2.7 cm, decurrent at base, serrate-toothed at margin, obtuse at apex, pubescent; 7-pairs; petiole sessile. Capitulum globose, solitary, terminal, 1.5 cm across; peduncle 4.5 cm long. Flowers ca. 3.5 mm long, pink. Female flowers and Maleflowers inner and outer; bract linear, ciliate at margin. Corolla tubular, 2.5 mm long, shortly 5-lobed, triangular. Stamens 5; filament slender, 1.5 mm long; anther lanceolate, 1.2 mm long. Ovary linear, 01 mm long; 1-loculed; ovules 1; style linear, 02 mm long; stigma shortly 2-fid or depply lobed. Achenes obovoid-oblong, glabrous and stalked; seeds 1, 1.2 mm long.

USES:

It is widely used in the treatment of skin diseases, jaundice, epilepsy, hemorrhoids, digestive disorders and asthma.

Sterculia urens Roxb. Sterculiaceae

Gum Karaya, Karaya, Senthanukku

Deciduous trees, to 15 m high, bark 10-12 mm, white or greenish-grey, smooth. Leaves simple, palmately 3-5 lobed, alternate, clustered at the tips ofbranchlets; petiole 8-18 cm long, stout, pubescent; lamina 15-30 × 18-30 cm, orbicular, base cordate or sagittate, lobe apex acuminate or caudate-acuminate, margin entire, glabrous above, velvety pubescent beneath, nerves 5-7 pairs, pinnate. Flowers polygamous, greenish-yellow, in axillary panicles, covered with sticky tomentose of glandular, stellate hairs. Calyx yellow, campanulate, hairy on both surfaces, lobes 5, a small hairy gland at the base of each lobe. Petals absent. Male flowers: stamina! column short with 10 anthers at its tip; bisexual flowers: carpels usually 5, free, superior, on a short stout gynophore; style short, thick, hairy; stigmas 5; stamens in a ring round the carpel. Fruit an aggregate of 4-6 follicles, red, densely pubescent, mixed with stinging hairs; seeds 3-6, brown or black, oblong.

USES:

Karaya gum is used as a bulk-forming laxative to treat constipation. It is also used for treating infections of throat, arthritis, joint pain and inflammation.

Strobilanthes ciliatus Wall.ex Nees Acanthaceae

Lesser Kurinj, Karvi, Karimkurunji

Shrubs, branches terete, glabrous. Leaves 10-15 × 3-5 cm, acuminate at both ends, glabrous, nerves 5 pairs, raised above, scabrid; petiole 2 cm long. Spikes 1.5 cm long, axillary, solitary, peduncled; bracts 6 × 4 mm, obovate; bracteoles 5 × 1 mm, linear-oblong. Flowers 1.5cm long; calyx 5 mm long, lobes oblong, connate below the middle; corolla 15 mm long, white with pink patches on lower lobes; stamina! sheath hairy. Capsules oblong, apically ciliate.

USES:

It is used to rheumatism, joint pain, skin diseases and respiratory issues like sinusitis, also used for coughs, fever, jaundice, urinary tract disorders, anti-diabetic and potentially anticancer.

Strychnos nux-vomica L. Loganiaceae

Nux vomica, Kuchla, Etti

A medium sized tree up to 10 m tall. Leaves simple, opposite, estipulate; petiole 5-13 mm, slender, glabrous; lamina 5-14 × 2-9 cm, elliptic, broadly elliptic or elliptic ovate, base acute to attenuate, apex acute or obtuse, margin entire, glabrous, shiny, coriaceous; 3-5 nerved from base, prominent, glabrous;. Flowers, greenish-white, in terminal sessile cymes, calyx lobes 4, 1 mm long, triangular, acute, pubescent; corolla greenish-yellow, tube 10 mm long, longer than lobes, slightly hairy near the base within, lobes 4, 4 × 1.5 mm, oblong, acute; stamens 4, anthers sessile inserted at the throat of corolla tube; ovary 2-celled, ovules many in each cell. Fruit a berry, 3-5 cm across, globose, yellowish-red or orange, glabrous; seeds 3 or 4, orbicular, flat, shiny, greenish-white, coin like, 2-2.5 cm diameter.

USES:

It is used for diseases of the digestive issues, nerve disorders, cardiovascular conditions, and arthritis. It is also used for insomnia, anxiety and gastrointestinal dis tress.

Strychnos potatorum L.f. Loganiaceae

Clearing Nut, Nirmali, Thethaan Kottai

It is a medium-sized, glabrous tree of height 12.5 m. Branches are swollen at nodes. Leaves are about 5-7.5 cm long, nearly sessile, sub-coriaceous, ovate or elliptic, acute, glabrous and shining, spuriously three or five nerved, with lateral nerves springing from the lower part of the mid rib, nearly reaching the tip. The base rounded or acute. Flowers are rather large, white, arranged in short, nearly sessile, axillary cymes. Fruit is a berry, black when ripe, globose, 1-2 cm in diameter, whitish, shining, with short adpressed yellow silky hairs. Seeds globose.

USES:

It is used in the treatment of gonorrhea, leucorrhea, kidney and bladder stones. It is also used in the treatment chronic diarrhea, dysentery and acts as a stomachic. Utilized for bronchitis, colds, liver ailments and epilepsy. The fiuit pulp is, considered as an effective emetic..

Switenia macrophylla G. King Meliaceae

Sky-fruit, Big-leaf Mahogany, Thenkani

Tall deciduous trees, to 25 m high; bark brown. Leaves paripinnate, alternate, estipulate; rachis 5-13 cm long, slender, glabrous; leaflets 4-10, opposite or subopposite; petiolule 3-4 mm long, slender, grooved above, glabrous; lamina 5.5-12 × 2.5-4.5 cm, obliquely ovate-lanceolate, oblong acuminate, oblong-lanceolate or falcate; base oblique, apex acuminate or caudate- acuminate, margin entire, glabrous, shining above, dull beneath, coriaceous; lateral nerves 7-12 pairs, pinnate, prominent, intercostae reticulate. Flowers, pale yellow, to 8 mm across, in axillary panicles to 10 cm; pedicel 2-4 mm. Calyx-lobes 5, lobes 0.5 mm, suborbicular. Petals 5, to 4 mm, oblong; disc annular. Stamina! tube 3.5-4 mm, apically IO-lobed; stamens 10. Ovary superior, 5-celled, ovules many; style to 1.5 mm long, attenuate; stigma some what lobed.. Capsule, 8.5 × 6.5 cm, oblong-globose, 5-valved, woody, and rusty out side; seeds many, to 6 cm, winged.

USES:

Seeds used to treat diabetes, hypertension, and improve blood circulation.. They have also been used to treat leishmaniasis and skin condition. Bark is used as an astringent for wounds and treating diarrhea, fever and to reduce inflammation.

Synedrella nodiflora (L.) Gaertn. Asteraceae

Cinderella weed, Mudianpacha, Mudinpachai

Erect dichotomous branched herbs; stem adpressed hairy. Leaves simle opposite, 4-8 × 2-4 cm, elliptic-ovate, base cuneate to truncate, narrowly winged on the petiole, margin serrate, apex acute, scabrous on both sides, basally 3-nerved; petiole to 3 cm. Heads solitary, axillary and terminal, few-flowered, 0.5-1 cm across, radiate. Phyllaries 2-seriate; outer foliaceous, inner paleaceous. Flowers heterogamous, yellow. Ray flowers: female; corolla tube 2-2.5 mm long; ligule 1-2 mm long, 2-3-lobed; ovary 3 mm long, oblong, winged; style-arms acute. Disc flowers; bisexual; corolla tube 2-3 mm long, lobes 5; stamens 5, included, ovary 2-3 mm long. Achenes 3-4 mm long, dimorphic: ofray florets dorsally compressed, with lacerate wings; of disc florets trigonous, not winged, with 2 or 3 rigid awns.

USES:

Leaves are rubbed as an antidote for Laportea crenulata allergy. Plant paste is applied for wounds, cuts and skin diseases.

Syzygium cumini (L.) Skeels (Myrtaceae)

Black Plum, Jamun, Naaval

Medium- sized evergreen tree, up to 12 m tall; bark scaly gray. Leaves opposite, elliptic-oblong, 14.5 × 3.5-6, cuneate at base, entire at margin, retuse-emarginate at apex, shiny, glabrous; nerves numerous; petiole 1.5 cm long. Panicles axillary-terminal, up to 4 cm long. Flowers ca. 04 mm across, white-cream; bracteoles small, deciduous, brown. Calyx-tube, petals 4, slightly lobed, rounded-cupshaped, after falling, glabrous. Stamens many, free, attached with calyx inwards; filament filiform, 05 mm long; anther 3-angled, 0.2 mm long. Ovary small, 2-celled; 2-loculed; ovules 1-2 per cell; style subulate, cylindric, 05 mm long; stigma small. Berries ovoid, ca. 2 cm across; seeds 1-2, globose, dark purplish-red, shiny.

USES:

Seed powder or bark decoction is much used for diabetes. Bark decoction is recommended for ulcers in the mouth and liver disorders. Seed decoction is consumed for frequent urination and fatigue. Fruit juice relieves excessive thirst, useful in curing dysentery. Tender shoot tip ground with milk or rice washed water is given to prevent abortion.

Taxus baccata L. Taxaceae

English yew, Binni

Dioecious trees 25 m tall and up to 4 m dbh. Crown normally pyramidal, becoming irregular with age; but many cultivated forms depart dramatically from this. Bark thin, scaly, brown. Leaves flat, arranged spirally but appearing 2-ranked, 10-40 × 2-3 mm, dark green. Pollen cones globose, 3-6 mm diameter, shedding pollen in early spring. Seed cone consists of a single seed surrounded by a soft, bright red aril 8-15 mm long. Arils mature gradually 6-9 months after pollination, and seeds are dispersed by birds

USES:

It has been used by the native populations for treating common cold, cough, fever, and pain. Recent years, the leaves and the bark is used in curing against breast, ovarian and Jung cancers.

Tephrosia purpurea (L.) Pers. Fabaceae

Wild Indigo, Sharapunkha, Kattukolingi

Erect/Decumbent undershrub, up to 70 high. Leaves imparipinnate; leflets opposite, 11-13, 5-7-pairs, elliptic-oblong to oblanceolate, 1.6-2.3 × 06-09 cm, cuneate at base, entire at margin, retuse, mucronate at apex, glabrescent; 9-pairs; petiole 02 mm long, villouse-pubescent. Racemes, up to 14 cm long. Flowers ca. 7 mm across, pinkish violet. Calyx-tube 02 mm long; lobe 5, equal, teeth, erect, hispid-pubescent. Petals 5; standard suborbicular, 07 mm across, acute at base, retuse at apex, densely hirsute pubescent; wings obovate-rhomboidea, 06 mm long; keels obovate, Falcate, 05 mm long. Stamens 10; filament subequal, flat; anther uniform, ovoid. Ovary flattened, 4.5 mm long; 2-loculed; ovules many; style flat, 03 mm long; stigma small. Pod flat 3.5 cm long, aristate at apex, apperssed-pubescent.

USES:

The plant paste is applied to heal cuts and wounds. Mastication of its roots gives strength to the teeth and gums. Root or whole plant decoction is taken for six months in case of chronic spleen disorders. Root juice mixed with honey is applied for pimples. The plant is used to treat jaundice, hemmroids and urinary disorders.

Terminalia arjuna (Roxb) Wight Combretaceae

Arjun, Arjuna/Kahu, Neer Maruthu

A large evergeen tree, up to 6-15 (-25) m tall with smooth, pale greenish to whitish grey bark. Leaves usually subopposite, oblong-elliptic or somewhat sub-orbicular, 7-18 (-25) cm long, 4-6 cm broad, obtuse, rarely subacute with rounded or cordate base, glabrous to subglabrous above, partially pubescent beneath, entire somewhat crenate or serrate in the upper half or throughout, petiole 5-10 mm long with 2 (-1) rounded glands at the apex. Inflorescence axillary or terminal paniculate spikes, 3-6 cm long, 1 cm broad. Bracteole small, deciduous, linear lanceolate. Flowers yellowish- white, sessile. Hypanthium broadly campanulate, 4-5 mm long, teeth triangular ca. 15 mm long, glabrous. Stamens much exserted. Ovary glabrous; disc barbate. Fruit ovoid-oblong, 2.5-5 cm long, brown with 5 projecting wings, thick narrow wings striated with 5 ascending veins.

USES:

It possesses the properties of strengthening the heart muscles thereby used to treat cardiovascular ailments. It has diuretic properties, reduces the chances of clot formation and helps in lowering blood lipid, thereby helpful in treating of high blood pressure.

Terminalia bellirica (Gaertn.) Roxb. Combretaceae

Belliric mycrobalan, Bahera/Bhuira, Thaandri

A large deciduous tree, 30 m tall. Leaves alternate or fascicled at the end of branches, elliptic or elliptic obovate, entire, acute or acuminate or rounded 8-20 × 7.5-15 cm, petiole 2.15 cm long, eglandular. Spikes axillary, loosely arranged, 5-15 cm long with pubescent floral axis. Flowers greenish- yellow, 5-6 mm across, sessile, upper flowers of the spike male, lower flowers bisexual; bracteoles minute caducous. Hypanthium 2-2.5 mm long, densly tomentose outside, densly villous with long brown hairs in side, expanded upward in 5 recurved, triangular calyx lobes, lobes ca. 1.5 mm long. Petals absent. Stamens 3-4 mm long, anthers ca. 1 mm long inserted on the hypanthium, epigynous. Disc intrastaminal, densly hairy. Style glabrous, fulvous, ca. 4 mm long, ovary ca. 2 mm long. Drupe obovoid 1.5-2.5 cm in diameter, covered with minute pale pubescence, stone very thick, indistinctly 5 angled.

USES:

Fruit is used to treat asthma, coughs, and bronchitis. It acts as a mild laxative and astringent to aid in constipation, diarrhea, and overall digestive health. It helps in treating skin diseases and promotes hair growth. It can improve insulin levels and reduce glucose levels. Seed extract with honey is taken for anti-ageing and managing arthritis. It is one of the ingredients of Triphala.

Terminalia chebula Retz. Combretaceae

Chebulic Myrobalan, Harar/Harad, Kadukaai

A medium size to large tree, up to 25 m tall with many spreading branches with pale greenish, gray and smooth bark. Leaves simple, altemate to subopposite, 10-15 × 3.75-7.5 cm, elliptic, -oblong or oblong, acute to obtuse, penninerved, covered with silky hairs when young, becoming glabrescent; rounded or cordate at base, usually unequal; petiole 1.25-3.5 cm long, pubescent, usually with 2, rarely more glands on the upper side at the apex. Spikes terminal, rarely panicled, 5-15 cm long with hispid floral axis. Flowers 5 mm across, yellowish; bracteoles linear, acute, pubescent, 5-6 mm long, caducous. Calyx tube 3-3.5 mm long, glabrous outside, hairy within, calyx teeth very short, broadly triangular. Petals absent. Stamens 10, in 2 series. Drupe ellipsoid or obovoid, obscurely 5- ribbed, 2-3.75 cm long.

USES:

Fruit decoction in hot water is givenn for constipation, It is also used to skin diseases, infections and to promote wound healing. Fruit rind decoction is recommended for immunity and support respiratory health.. It is one of the ingredient of Triphala, often referred as King of medicines.

Thespesia populnea (L.) Soland. ex Correa Malvaceae

Portia tree, Paras Pipal, Poovarasu

Small tree upto 7.5m tall; young parts covered with brown, peltate scales. upto 10 m tall. Leaves alternate, broadly ovate, cordate, 8-13 × 6-10, cordate at base, entire at margin, acuminate at apex, glabrous; 7-pairs; petiole 5-7 cm long. Stipule linear-subulate. Flowers solitary, axillary. ca. 6 cm across, yellow with purplish center, solitary, axillary; pedicel 5 cm long. Epiclayx segment 3-5, ca. 05 mm long. Calyx-lobe 5, tube cylindric, 1.5 cm long, lobe minute teeth. Petals 5, obovate 7 cm long. Staminal sheath, 2.5 cm long. Stamens many; filament filiform, 03 mm long. Ovary globose-ovoid, ca. 08 mm long; 5- loculed; ovules 3 per cell; style filiform, ca. 5 cm long; stigma 5, clup-shaped. Capsules globose, 2.5 cm across, depressed, glabrescent with yellow sap; seeds 2-4- per cell, obovoid.

USES:

Twigs are used as tooth brush in case of toothache. Bark decoction is recommended for skin diseases. Fruits and bark decoction is much used to wash for septic wounds and ulcers. Bark decoction is dysentery and diabetes.

Tinospora cordifolia (Willd.) Miers Menispermaceae

Tinospora, Gurch, Seendhal

It is climbing shrub with heart-shaped leaves. It has stems about 6 cm in diameter, with light grey, papery bark. The leaves are 7.5-14 cm long, 9-17 cm broad, broadly ovate or orbicular, deeply heart shaped at the base. Tiny greenish yellow flowers occur in racemes 7-14 cm long. Flowers have 3+3 sepals in 2 layers, the outer ones are small, the inner large large. Six stamens prominently protrude out. The plant flowers during the summer and fruits during the winter.

USES:

Whole plant decoction is used to boost immunity, treat chronic ulcers, and reduce inflammation. It acts as an adaptogen to manage stress supports liver health, often used for allergies, respirattory issues. It also helps in regulate blood sugar levels improves insulin sensitivity and is used to treat metabolic disorders.

Trachyspermum ammi L. Apiaceae

Ajowan caraway, Ajwain, Omamam

An annual, erect, aromatic herb, 75 cm tall, stems striate, glabrous, usually strongly branched. Leaves alternate, pinnately compound; petiole long, sheathing; blade in outline ovate-elliptical, up to 13 cm × 12 cm, 2-3-pinnate, segments linear to narrowly oblong, up to 2 mm long. Flowers in compound umbel, up to 6 cm in diameter. calyx teeth 0.5 mm long, fleshy, petals obcordate, 0.6-0.7 mm long, apex inflexed, white; stamens 5, radiating, anthers reddish-brown, ovary inferior, densely white hairy, styles 2, stigma globose. Fruit a flattened, subglobose schizocarp, splitting into 2 hairy, I-seeded mericarps, 2 mm × 1 mm, each mericarp with 5 longitudinal ribs, on ribs with broad, warty trichomes, each mericarp with 4-6 oil ducts.

USES:

The seeds are used to cure digestive disorders, and indigestion, flatulence, dianhea and abdominal pain. Also serves as an expectorant, To reduce mucus easing bronchitis and asthma It is often used for cough relief. Chewing seeds in gargles helps alleviate toothaches and reduce oral infections. It is used to regulate menstrual flow, relieve cramps and sometimes to assist with postpartum recovery. Used in removing kidney stones, enhance appetite and reduce obesity.

Tragia involucrata L. Euphorbiaceae

Indian Stinging nettle, Pit Pami, Kanchori

Slender, twining herbs with stinging hairs. Leaves 6-10 × 3-5.5 cm, ovate or elliptic, base acute or rounded, margin serrate, apex acuminate, hispidulous on both sides; petiole to 2.5 cm long. Spikes axillary, monoecious, to 2 cm long; male flowers above, female flowers 1-2, at the base. Male flowers ca. 1.5 mm across; bracts spathulate; tepals 3, spreading; stamens 3, anthers subsessile. Female flowers ca. 3 mm across, ebracteate; tepals 6, ca. 1 mm long, ovate-lanceolate, enlarged and spreading in fruits; style 3, spreading. Capsule ca. 0.6 × 1 cm, 3-lobed, hispid. Seeds globose.

USES:

The decoction prepared by the roots is given m the diseases like fever, bronchial ailments, cough. The roots of this plant along with Turmeric and Tulsi is made into paste and applied over skin infections like pruritis scabies. The paste of fruits is applied for excessive hair loss.

Trewia nudiflora L. Euphorbiaceae

False white Teak, Gambhar, Aattharasu

Deciduous dioecious trees to 20 m, bark greyish-brown, smooth; blaze creamy yellow; young parts tomentose. Leaves simple, opposite, decussate; stipules inter petiolar, 2-3 × 1 mm, linear lanceolate; petiole 3-12 cm long, slender, glabrous, unequal at each node, lamina 6-20 × 9-15 cm, broadly ovate or deltoid base subcuneate, subtruncate, truncate, subcordate or cordate at base, apex acuminate to cuspidate or cordate-acuminate, margin entire, glabrous above, glaucous beneath, chartaceous; 3-ribbed from base, prominent, lateral nerves 3-6 pairs, prominent, intercostae scalariform, faint. Flowers unisexual, pale green; male flowers; 3-5 mm across, numerous on 10-20 cm long peduncle with abundant floral remains over the peduncle; bracts ca. 3 mm across, ovate-lanceolate, pubescent; pedicels 4-5 mm long; tepals globose, ovate, splitting into 3-4 broad, concave, often reflexed segments, 4-7 × 3-4 mm; stamens many, clustered on a convex receptacle; filaments 1.5-3 mm long, free, slender; anthers oblong; female flowers; 5-9 mm across, 4-5 on axillary racemes; pedicels 5-10 mm long, finely hairy; tepals 3-5, 4-5 mm long, broadly ovate, densely tomentose beneath, cauducous; ovary superior, 3-4 mm across, 2-4-loculed, ovule one in each cell; styles 3, 1.5-3 cm long, spreading, connate below, oblong. Fruit a capsule 2-3 cm across, woody, dark brown, broadly rounded, 3-4 loculed, pericarp very thick; seeds globose, ovoid.

USES:

The plant has potential activity again st ca n cer, tuberculosis, anti-inflammatory, antiulcer and antimicrobial activities.

Trianthema portulacastrum L. Aizoaceae

Desert Horse-Purslane, Lalsabuni, Sarannai

Prostrate or ascending succulent herbs; branchlets thickened at the nodes. Leaves opposite or subopposite, unequal, 1.2-3.5 × 0.8-2.2 cm, orbicular-obovate, oblong or elliptic, base obtuse or attenuate, margins entire, often undulate, purplish, apex obtuse, rounded or retuse, sub-succulent; Flowers ca. 7 mm across, axillary, solitary, sessile, sheathed by the base of the petiole. Calyx tube ca. 1.5 mm long, adnate to base of petiole; lobes 5, white or pink, ca. 4 mm long, oblong, obtuse. Stamens 15-20. Capsule to 5 mm long, turbinate, apex truncate, 2-lobed, brown, circumscissile.

USES:

It is used as analgesic, diuretic, laxative and used in the treatment of jaundice. The fleshy leaves are applies as a poultice to heal thermal wounds, showing improved collagenation and skin tensile strength. It also Serves as an antidote for alcohol poisoning and as acathaartic and abortifacient

Tribulus terrestris L. Zygophyllaceae

Devil's thorn, Gokhru, Nerunjil

Prostrate silky herbs. Leaves opposite, imparipinnate; leaflets opposite, 5-7 pairs, 1.5 × 0.7 cm, oblong; petiole to 7 mm, petiolule 1 mm, stipules 2-3 in cluster, persistent. Flowers axillary, 1 or 2, golden yellow, 2 cm across; pedicels to 3 cm, ; sepals 5, 5 mm, linear; petals 5, 7 × 4 mm; stamens 10, filaments dilated at base, disc IO-lobed; ovary 5- lobed, hirsute with bulbous hairs; stigmas 5. Fruit 1 cm across, globose, 5-angled, cocci 5, woody, each with a pair of divaricate spines.

USES:

Seed or fruit decoction is used for urinary tract infections, oedema and urinary stones. Seeds decoction is also recommended for urine block and habitual abortion. It is a tonic and has aphrodisiac property. Fruit or plant ash is applied for swelling and rheumatism. It is also used in treating blood pressure, inflammation, pain and skin conditions.

Trichosanthes cucumerina L. Cucurbitaceae

Snake gourd, Jangli chachinda, Paipoodal

Monoecious annual herb, climbing by 2-3-branched tendrils; stem slender, 5- angled. Leaves alternate, simple; stipules absent; petiole 2-10 cm long, furrowed, succulent, scabrid hairy; blade slightly to deeply 5-7-lobed, 7-25 cm × 8-20 cm, cordate at base, margin dentate, pubescent. Flowers unisexual, regular, 5-merous, white. Calyx tubular. Corolla lobes fringed with hair-like outgrowths; male flowers in 5-many-flowered axillary racemes on 10--30 cm long peduncles, with 3 stamens; female flowers solitary and sessile, with inferior, I-celled ovary, long-hairy, stigmas 3. Fruit a very slender, long and cylindrical berry, often twisted, 30--180 cm × 2-10 cm, greenish-white when immature, dark red when mature, many-seeded. Seeds flattened, 1-1.5 cm long, greyish-brown, sculptured, margin undulate.

USES:

It is used in the treatment of diabetes, pain and reduce inflammation. It is also employed in curing bronchitis and soothe throat irritation. Tropical applications of fruit paste to treat eczema, dermatitis and psoriasis.

Trigonella foenum-graceum L. Fabaceae

Fenugreek, methi, Vendhayam

Erect, stiff, annual herb with a well-developed taproot, often reaching 30-45cm in height. Leaves alternate, trifoliolate (three-leaflet) leaves that are grey-green and obovate in shape. Flowers small, sessile, and pale yellow to whitish in color, blooming from January to March. The plant produces long, slender (5.7-8 cm), curved pods with a persistent beak, containing 10-20 yellow-brown, hard, furrowed seeds.

USES:

It is used in curing diabetes, serves as a demulcent, known to aid in treating obesity, boosting testosterone levels and improving libido.

Tylophora indica (Burm.f.) Merr. Asclepiadaceae

Indian ipecac, Antamul, Nanjuaruppan

A slender, twining or creeping perennial herb with watery sap. Leaves opposite, ovate or ovate-oblong 5-10 cm long, acute or acuminate at the tip, rounded at the base, with a smooth, glabrous upper surface and sometimes slightly hairy beneath. Flowers small, greenish-yellow or greenish-purple, arranged in axillary umbellate cymes. Fruits a follicle, generally in pairs, tapering at the apex, and glabrous. Seeds small, flat, and ovate with a silky white coma.

USES:

This plant is having anti-asthmatic, expectorant and anti-rheumatic properties. Roots are used in chronic bronchitis, asthma, early stages of whooping cough and relieving pain due to gout. The powdered root is used for treating intermittent fever and root paste is used in jaundice. The decoction of leaves or root is useful for treating in joint pain, rheumatoid arthritis and inflammatory skin conditions.

Vernonia cinerea (L.) Less. Asteraceae

Little ironweed, Sahadevi, Puvamkuruntal

Erect herb up to 75 cm tall, woody at base; stem much branched, terete, much branched terete, ribbed pubescent. Leaves simple, ovate-elliptic to oblanceolate, 1.5-4 × l-3 cm, sometimes orbicular, rounded or acute at apex, subentire or irregularly dentate, base narrowed into the petiole, sparsely hairy above more so below. Heads up to 0.7 cm across homogamous, in terminal corymbs. Involucral bracts, linear-lanceolate, pubescent. Florets pink or lilac. Corolla up to 0.4 cm long, lobes 5, hairy, tubular campanulate. Stamens 5. Achenes obscurely 4-5 angular, finely ribbed, liner -oblong, dark brown, with white pappus hairs.

USES:

Oil is prepared from plant juice is used as hair oil (has cooling effect), as it helps in the proper growth and colour of the hair, also prevents hair fall. Whole plant or root decoction is given for fever. It also serving as a natural sedative and potential aid for smoking cessation.

Vetiveria zizanioides (L.) Nash. Poaceae

Vettiver, K.has-K.has, Vettiver

Densely tufted perennial grass. Leaves linear-lanceolate, folded towards base; sheaths compressed. Flowers in panciles; branches verticellate, spreading at antheis. Sessile spikelets; lanceolate, bisexual, with a shortly bearded callus. Glumes equal, coriaceous; lower spinulose; upper boat-shaped, keeled, ciliate. Lower lemma linear, hyaline, keel winged; upper oblong, hyaline, 2-lobed, awned; palea linear, hyaline, neverless. Stamens 3. Pedicelled spikeles, lanceolate. Male: Glumes equal, coriaceous; lower lanceolate, spinulose; upper boat-shaped, keeled, ciliate; lower lemma lanceolate; upper lemma oblong, 2-lobed. Stamens 3. Fruits oblong, caryopsis.

USES:

Root decoction is a blood purifier. Its decoction with Scoparia dulcis and Cyperus rotundus is prescribed for shivering during fever. Syrup prepared from its roots has cooling effect, causes sweating and is used for fever. Root decoction relieves thirst, causes perspiration, increases urine and purifies blood. Root paste is applied over stomach for preventing miscarriage in pregnant women.

Vitex negundo L. Verbenaceae

Indian privet, Nirgandi, Nochi

Shrubs or small trees; purple pubescent all over, aromatic, bark pale. Leaves 3-5- foliolate; leaflets 6-13 × 2-5 cm, narrowly oblong or elliptic to lanceolate, base acute, apex acuminate. Panicles terminal, 10-25 cm long. Calyx 5-toothed obconic, ca. 3 mm long, teeth triangular. Corolla deep purple to violet in colour, ca. 7 mm across, hypocrateriform; tube 3-5 mm long, puberulent without, upper lipd 2-lobed, lower 3-lobed with the middle lobe larger, obovate, undulate-margined, other lobes shorter, subequal, obtuse. Stamens 4, filaments purple. Ovary ca. 1 mm long; style purple; stigma 2-fid. Drupe 3-5 mm across, globose, purple or black.

USES:

The leaves are used in curing rheumatoid arthritis, osteoarthritis, sciatica and sprains. It also helps in managing bronchitis, asthma and coughs. It is widely employed in managing menstrual irregularities and uterine disorders. Leaf juice is applied on scalp for proper hair growth.

Wattakaka volubilis (L.f.) Stapf Asclepiadaceae

Green Milkweed Climber, Harandodi, Kurinjan

Large twining shrubs. Leaves simple, broadly ovate, cordate at base. Flowers green, in lateral umbellate cymes. Calyx 5-lobed, divided to the base. Corolla rotate; lobes triangular. Corona single, of 5 fleshy top-shaped segments attached to the upper part of the starninal column. Pollinia solitary in each anther-sac. Fruits ovoid-oblong follicle, tomentose when young, tapering to a blunt apex.

USES:

Leaf paste is used for boils and ulcers in cattle. This paste is used to swelling and pain, including rheumatic pain and skin inflammation. Leaf juice is used for diabetes. Leaf paste with lime juice is applied internally for sore throat. Five leaves ground in milk are given for three days from 5th day of menses for curing infertility due to anomaly in egg formation.

Withania somnifera (L.) Dunal Solanaceae

Asvagandha, Asgandh/Ashwagandha, Amukkara

A suffruticose shrub, 60-90 cm tall. Branches ascending. Shoots stellate tomentose. Leaves 4-13 × 2-9 cm, elliptic-ovate to broadly ovate, acute, cuneate or oblique, entire to repand. Petiole 10-20 mm long. Flowers sessile to subsessile, greenish yellow, in axillary clusters of 2-5. Calyx 2-2.5 × 2.5-3 mm, up to 2.0 × 14 mm in fruit, stellately hairy, becoming glabrescent and membranous; teeth 1-1.5 mm long, up to 8 mm in fruit, acute. Corolla lobes 2-2.5 mm long, triangular, tomentose to the outside. Anthers sub included; filaments 1-2.5 mm long. Berry globose, 6-8 mm broad, red. Seeds sub pyriform to reniform, minutely reticulate-foveolate, yellowish-brown.

USES:

This plant roots are used reducing cortiserol levels and calms the nervous system. Serves as a natural sedative and improves sleep quality, aids in improving memory and acts as a cognitive booster. It helps in enhancing muscle mass endurance and overall physical strength. It also used in lowering blood sugar levels and stimulates insulin secretion. It is widely employed in reducing joint inflammation and acting as an antiarthritic

Wrightia tinctoria (Roxb.) R. Br. Apocynaceae

Pala Indigo Plant, Dudhi/Khirni, Veppaalai

Deciduous trees upto 15 m; Leaves simple, opposite, distichous, estipulate; petiole 2-5 mm long, stout, glabrous; lamina 3.5-11 × 2-4 cm, oblong-lanceolate or elliptic ovate; base acute; apex acuminate or caudate acuminate; margin entire, nerves 5-10 pairs, slender, arched, pinnate. Flower white, scented, in terminal cymes. Pedicel 1.5 mm long, bracts 2, minute. Calyx lobes 5, with membranous margin, 2 mm, ovate, obtuse, ciliate, glandular inside. Corolla salver shaped, lobes 5, broad lobes 12 × 5 mm, oblong, obtuse, throat with 1-2 series of erect, fimbriate corona scales, tube 1.5-2 mm. Stamens 5, insrerted on the mouth of the tube, anthers sagittate, spurred at base, conniving and adhering to the stigma; disc absent. Carpels 2, free, many ovuled, style filiform, stigma ovoid, usually with a toothed basal ring. Fruit of 2 follicular mericarps, 15-45 cm long, slender, smooth, green, cylindric, cohering at the tip; seed 16 mm long, black, commate at one end.

USES:

Leaf paste is used to treat toothaches and fillcavities. Leaf paste in coconut oil is applied for skin diseases. Leaf extract or paste in virgin coconut oil is applied for psoriasis, eczema, scabies and dandruff Seed powder used in curing diarrhea and sexual debility. Its leaves and bark are used in curing jaundice and stomach ailments.

Zingiber officinale L. Zingiberaceae

Ginger, Adrak, Inji

It is a biennial or perennial, creeping rhizome, and an annual stem, which rises 30 45 cm in height, is solid, cylindrical, erect, and enclosed in an imbricated membranous sheathing. The leaves are lanceolate, acute, smooth, 12 or 15 cm long by about 3cin breadth, and stand alternately on the sheaths of the stem. The flower-stalk rises by the side of the stem from six inches to a foot, and, it, is clothed with oval acuminate sheaths; but it is without leaves, and terminates in an oval, obtuse bract, imbricated spike. The flowers are of a dingy yellow color, and appear two or three at a time between the bracteal scales.

USES:

It helps in reducing gas and improving digestion, relieving nausea, easing a cold or the flu, relieving pain, reducing inflammation, supporting cardiovascular health and lowering cancer risk.

Ziziphus mauritiana Lam. Rhamnaceae

French Jujube, Ber/ Baer, Elandai

Trees, to 10 m high with drooping branches and tomentose branchlets; branches armed with curved spines. Leaves simple, ovate or orbicular, dark green ad shining above, white tomentose below petiole 4-8 mm, stout; lamina 1.5-4 × 1.5-3 cm, ovate, elliptic-ovate, elliptic-orbicular or suborbicular, base oblique, subcordate or round, apex round and retuse, strongly 3-ribbed from base, convergent; lateral nerves 2-3 pairs, pinnate, prominent. Flowers 5 mm across, greenish-yellow, in 15-20 flowered dense cymose axillary fascicles. Peduncle much reduced. Calyx-tube 0.5 mm, woolly outside, glabrous within; lobes 5, triangular, prominently keeled on inner face. Petals 5, cucullate, round at apex, deflexed with the stamens; disc 10-lobed, grooved. Stamens 5, enclosed in the petals, not exerted beyond them; filaments slightly dilated at base. Ovary embedded in disc, 2-celled; ovule 1, ascending; styles 2, connate to the middle; stigma 2-fid. Fruit a drupe 1 × 0.5 cm, oblong-globose, yellow or orange when ripe.

USES:

It is used to treat anxiety, insomnia, diarrhea, skin infections, and inflammatory conditions. Its leaves, fruits, bark, and roots are having hepatoprotective properties, often used to aid digestion, treat fevers, and accelerate wound healing.

Ziziphus xylopyrus (Retz.) Willd. Rhamnaceae

Woody-fruited jujube, Ghat bor, Kottai Elanthai

It is a thorny evergreen shrub up to 4 meters. Leaves are alternate, entire oblique, rounded symmetrical base, and obtuse at apex, serrulate margin, and about 2-7 cm long. Flowers are small, yellowish or yellowish white, 4-6 cm across. Calyx-lobes 2.0-2.5 mm long, keeled up to the middle, glabrous inside pubescent outside. Five petals, 1.5-2.0 mm long, obovate. Stamens five, disc 10-60 lobed, rarely 5-lobed, and glabrous. Berry, globular or round, 1.2 to 1.8 cm across.

USES:

This plant used to treat diarrhea, digestive disorders, skin infections, and fever. Its bark, fruits, and leaves contain high tannin content, providing astringent, anti-inflammatory, and wound-healing properties, and are used for pyorrhea and respiratory relief.

www.ingramcontent.com/pod-product-compliance
Lightning Source LLC
LaVergne TN
LVHW070946210926
848482LV00002B/111